Additional

THE CA

"Thomas R. Cech led the way into what will be known as the age of RNA. When others were focused on DNA, Cech probed the mysteries of the then lesser-known, but wondrous molecule that creates the stuff of life and is a key to life's origins. The Catalyst vividly describes the marvels of RNA and the discoveries—from vaccines to gene-editing tools—that will shape our future. I am so glad Cech wrote this book."
—Walter Isaacson, #1 New York Times best-selling author of Elon Musk and The Code Breaker

"Nobel laureate Thomas R. Cech takes us into the world of RNA with a story that's as enlightening as it is fascinating. It's a must-read for anyone interested in the molecule that has shaped life itself and is driving the future of science and medicine." —Jennifer Doudna, Nobel laureate, co-inventor of CRISPR gene editing, and founder of the Innovative Genomics Institute

"[An] impressively comprehensive work . . . [Cech] writes with easy charm and uses analogies helpfully and elegantly."
—Nessa Carey, Times Literary Supplement (UK)

"A series of dogma-smashing and Nobel-winning discoveries about RNA has transformed our understanding of how life works and given birth to exquisite, life-saving technologies. . . . The Catalyst is a masterful account of the RNA revolution in biology and medicine by one of its brilliant pioneers." —Sean B. Carroll, author of The Serengeti Rules and A Series of Fortunate Events

"For brains of a mortal frequency, RNA may have only popped into consciousness with the groundbreaking introduction of COVID-targeting mRNA vaccines and CRISPR therapies. In the telling of Nobel Prize–winning scientist Tom Cech, RNA was long the sidelined brother

of DNA, with its showy double helix. Thanks to work by Cech and others, the power of RNA to impact aging and catalyze biochemical reactions is made apparent."
—Literary Hub

"An absorbing account of the scientific journey that led to the development of crucial RNA-based vaccine technologies and other important medical breakthroughs. Readers will come away from this fascinating book feeling more knowledgeable and empowered to make decisions about their own health."
—Kenneth Frazier, retired CEO of Merck & Co.

"Cech debuts with an entrancing primer on 'the vast capabilities of RNA.' . . . The biological discussions are remarkably lucid, thanks to easy-to-understand analogies. . . . [The Catalyst] fascinates."
—Publishers Weekly, starred review

"Cech is a lucid prose stylist, vividly communicating his and his colleagues' excitement as they have unraveled RNA's secrets. . . . [The Catalyst is] an expert update on the hottest topics in biology."
—Kirkus Reviews

THE CATALYST

The
CATALYST

RNA AND THE QUEST
TO UNLOCK LIFE'S
DEEPEST SECRETS

Thomas R. Cech

W. W. Norton & Company
Independent Publishers Since 1923

Copyright © 2024 by Thomas R. Cech

All rights reserved
Printed in the United States of America
First published as a Norton paperback 2025

For information about permission to reproduce selections from this book, write to Permissions, W. W. Norton & Company, Inc., 500 Fifth Avenue, New York, NY 10110

For information about special discounts for bulk purchases, please contact W. W. Norton Special Sales at specialsales@wwnorton.com or 800-233-4830

Manufacturing by Lakeside Book Company
Book design by Brooke Koven
Production manager: Julia Druskin

ISBN 978-1-324-11087-3 pbk.

W. W. Norton & Company, Inc.
500 Fifth Avenue, New York, N.Y. 10110
www.wwnorton.com

W. W. Norton & Company Ltd.
15 Carlisle Street, London W1D 3BS

10 9 8 7 6 5 4 3 2 1

*To my parents, Robert and Annette,
who fostered my love of science,
and to Carol, Allison, and Jennifer,
who've accompanied me on the journey.*

CONTENTS

Introduction: **The Age of RNA** 1

PART I: **THE SEARCH**

1. **The Messenger** 9
2. **Splice of Life** 29
3. **Going It Alone** 47
4. **The Shape of a Shapeshifter** 69
5. **The Mothership** 91
6. **Origins** 109

PART II: **THE CURE**

7. **Is the Fountain of Youth a Death Trap?** 129
8. **As the Worm Turns** 153
9. **Precise Parasites, Sloppy Copies** 169
10. **RNA versus RNA** 183
11. **Running with Scissors** 205

Epilogue: **The Future of RNA** 231

Acknowledgments 239
Glossary 241
Notes 251
Index 281

THE CATALYST

Introduction

THE AGE OF RNA

It is often said that the first half of the twentieth century was the age of physics. The curvature of space-time, the dynamics of subatomic particles, the Big Bang and black holes, the unleashing of atomic energy that could power whole cities or obliterate them: all these discoveries burst forth to revolutionize science and change our everyday lives. One could say it was the big bang of physics itself, taking place roughly between 1905, when Einstein gave us $E = mc^2$, and 1947, when the transistor was invented at Bell Labs.

As the second half of the twentieth century dawned, biology began to nudge physics out of the scientific spotlight—and by "biology," I mean DNA. After all, that half century more or less began with Francis Crick and James Watson's momentous discovery of the DNA double helix in 1953, and it ended with the Human Genome Project (1990–2003), which decoded all of our DNA into a biological blueprint of humankind. Today, everyone knows DNA's greatest hits: that it contains our genetic information, that it can be used to trace our ancestry and pinpoint hereditary diseases and solve crimes. DNA has even entered the vernacular. If I tell you that something is "in my DNA"—whether it's a devotion to hiking in the mountains or a love of Thai food—I'm saying that it's a core part of who I am, my essence.

During the age of DNA, RNA was mostly overlooked by the general public. Certainly, textbooks described, and students learned, how RNA—the ribonucleic acid to DNA's deoxyribonucleic acid—was copied from the double helix and how *messenger RNA* (mRNA) worked to transmit DNA's code to instruct the synthesis of proteins. But RNA was never the star of the show. It was like a biochemical backup singer, slaving away in the shadow of the diva.

Yet, to those in the scientific community, RNA's previously unappreciated talents began to become apparent. RNA is tiny, measuring only about a nanometer in diameter. If you stacked *molecules* of messenger RNA side by side, you could fit 50,000 of them within the breadth of a single human hair. Yet, as researchers started to discover, what RNA lacks in size it makes up for in versatility: folding into origami-like shapes, it can pull off wild stunts that make its genetic parent, DNA, look like a one-trick pony.

In fact, DNA has just one trick, albeit one that's central to all life on Earth. DNA stores genetic information. That's it. It's like the hieroglyphics in an Egyptian mummy's tomb, or the grooves in a vinyl LP record, or the ones and zeros that make up the bits of information stored in a computer. DNA's job is to sit there, in the cell nucleus, storing information. Reading out that information and doing something with it requires proteins—and RNA.

The first thing to understand about RNA, then, is that it is a many-splendored thing. Yes, it can store information, just like DNA. As a case in point, many of the viruses that plague us don't bother with DNA at all; their genes are made of RNA, which suits them perfectly well. But storing information is only the first chapter in RNA's playbook. Unlike DNA, RNA plays numerous active roles in living cells. It can act as an *enzyme*, splicing and dicing other RNA molecules or assembling proteins—the stuff of which all life is built—from *amino acid* building blocks. It keeps stem cells active and forestalls the aging process by building out the DNA at the ends of our *chromosomes*. By guiding the gene-editing machinery of *CRISPR*, it empowers us to rewrite the code of life. Many scientists believe RNA even holds the secret of how life on our planet began.

At last, RNA has begun to step out of DNA's shadow to reveal its own immense potential. Since 2000, RNA-related breakthroughs have led to 7 Nobel Prizes. In that same period, the number of scientific journal articles and the number of patents generated annually by RNA research have each quadrupled. There are more than 400 RNA-based drugs in some stage of development, beyond the ones that are already in use. And in 2022 alone, more than $1 billion in private equity funds was invested in biotechnology start-ups to explore new frontiers in RNA research.

While DNA may have dominated biology research in the past, RNA has clearly become the focus of the future. The twenty-first century is already standing out as the age of RNA—and this century still has a long way to go.

This book is an educated citizen's guide to understanding how RNA—literally and metaphorically—went viral, how it transformed from an arcane topic mostly of interest to biochemists to a mainstream subject that is shaping the future of science and medicine.

I come to the story not as an impartial observer but as an active participant. As a chemistry and biochemistry professor at the University of Colorado in Boulder, I have been studying RNA for most of my career. I've been an eyewitness to discoveries about RNA that caused scientists to rethink the deep question of how life originated on planet Earth and revealed breathtaking insights about human health and disease. A few of these discoveries my research group and I made ourselves. Others were the work of close friends and colleagues—and so it only feels right to refer to them by their first names.

Taken together, these breakthroughs in RNA research represent one of the most transformative scientific achievements since the discovery of the DNA double helix. Yet, for many years, the public was not well equipped to appreciate this accomplishment because people had only a hazy idea of what RNA is or why scientists were so excited about it. I always thought this was a shame, for the stories are thrilling. In addition, the public financed much of this research with its tax dollars, and people deserve to know how their investment has paid off.

Then, in the tumultuous spring of 2020, the public came face to face with RNA in a striking way. My work, like that of so many others, had for the moment come to a halt. My lab was shut down, my classes were canceled. But my subject was suddenly on the tip of everyone's tongue. The world was being ravaged by SARS-CoV-2, the RNA-based virus that causes Covid-19. To combat the virus, *mRNA vaccines* were being developed with unprecedented speed, a stunning achievement built on decades of fundamental research breakthroughs in RNA science—breakthroughs about which most people knew nothing at all.

Naturally, the public wanted to understand this molecule that was both the cause of and the potential cure for our troubled times. So I went from RNA scientist to RNA spokesperson. I made it my mission to demystify RNA, first in public talks and now with the book you hold in your hands.

I tell the story of RNA in two parts. The first is the story of how RNA revealed itself as life's great catalyst. We begin in the 1950s with the experiments that uncovered how RNA orchestrates the construction of the proteins that perform most of the essential functions in living organisms, from holding cells together to metabolizing food. Then we see how RNA, through a curious transformation called *splicing*, is responsible for helping us humans perform so much more with our DNA information than, say, a fungus, a worm, or a fly can.

From there, the story takes a personal turn. I recount how my team discovered catalytic RNAs called *ribozymes*, whose existence violated what had been considered to be a bedrock rule of nature—that enzymes must be proteins. This breakthrough led to the 1989 Nobel Prize in Chemistry and marked a major turning point in RNA's story—the moment when the world of science started to see this molecule not as a passive messenger, a bit player in the chemistry of life, but as the star of the show.

The next major challenge was to map the wondrous shapes that RNA assumes to accomplish its multiple miracles (an effort that had humbled even the great James Watson after his success solving the structure of DNA). We then observe how RNA was discovered to be the secret power source behind the *ribosome*, the "mothership" in our cells that reads the code contained in messenger RNAs and then uses it to construct the proteins that power so much of life. Finally, we examine how RNA could answer the greatest chicken-or-egg problem in science: how life on Earth began almost 4 billion years ago.

While the first part of the book describes how RNA supports life, the second part shows how RNA can improve and extend life beyond nature's current limits. We start with the extraordinary story of *telomerase*, an RNA-powered enzyme that has taught us that immortality and cancer are really flip sides of the same coin. Next, we learn how tiny RNAs that work like switches, turning off messenger RNAs in cells, are being repurposed to short-circuit disease.

But RNA doesn't only cure us; it also can kill us. RNA is the genetic material of many of the deadliest viruses in history, from polio to SARS-CoV-2. While these viruses make RNA into a villain, mRNA vaccines show us how RNA can still save the day, protecting us not only from Covid-19 but also, perhaps, from cancer and many other diseases.

And, at last, RNA gets its ultimate revenge on DNA as the power behind CRISPR, a system that gives us the ability to remodel DNA itself. CRISPR has already revolutionized fundamental scientific research, and profound applications in medicine and in slowing the march of climate change are almost upon us. As it turns out, the same versatility that enables so many vital functions in nature also makes RNA the perfect tool for biomedical engineers who are now redefining life as we know it.

Given that the hero of this book is present in every living thing on the planet and has been around for about 4 billion years, there was no way I could tell the whole, unabridged story of RNA. I had to make hard choices about what to include and what to leave out. I also had to simplify many of the scientific concepts to make them more accessible. At times I describe RNA as spaghetti and compare RNA splicing reactions to copying and pasting text in a word-processing program. Such simplification may irritate my colleagues, but—as my wife (and fellow biochemist) often reminds me—I am not writing this book for them.

I expect I'll annoy my science friends in yet another way. In telling this tale, I have tried to keep the spotlight on RNA itself. So while my stories feature some of the researchers who stumbled onto key discoveries or who took several wrong turns before uncovering the truth, I make no claims at being comprehensive. For each scientific topic covered in this book, I have tried to credit additional researchers in the endnotes, although I apologize in advance for the multiple offenses of omission I've likely committed. One of the wonderful things about today's RNA science is that most discoveries have multiple contributors who are constantly passing the ball to each other as in a game of rugby. Also as in rugby, the researchers occasionally pile into a ruck, where for a while the ball seems lost. It's a competitive, messy, at times painful game, but one that is punctuated by moments of glory.

PART I

THE SEARCH

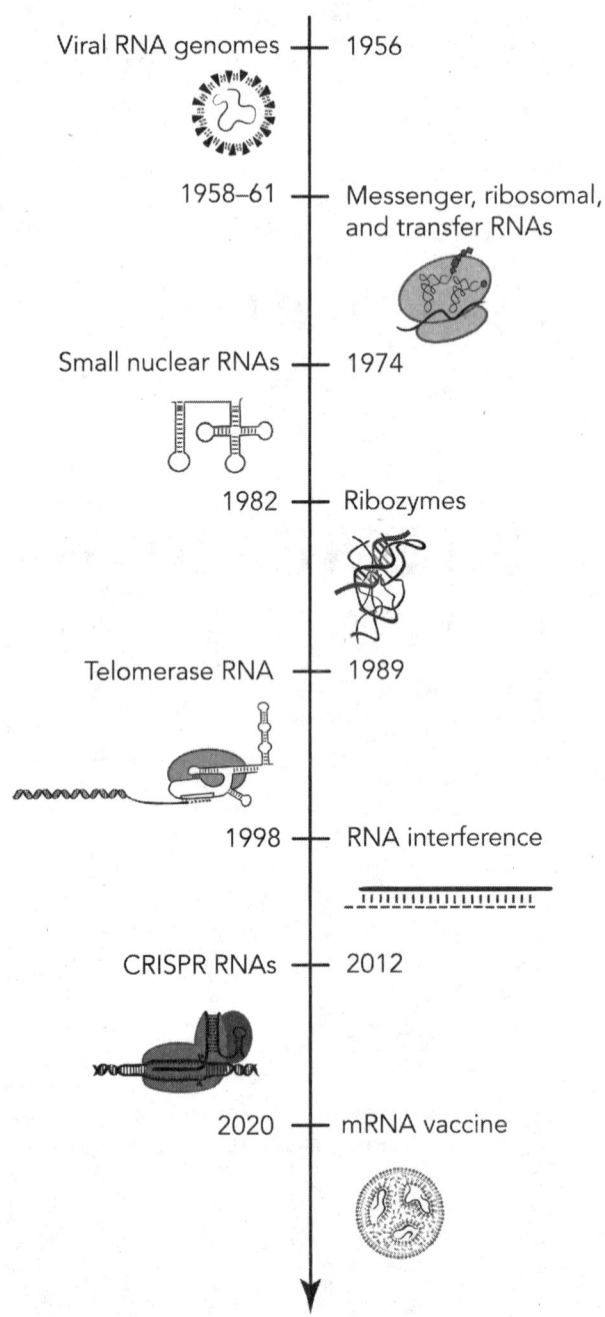

Chapter 1

THE MESSENGER

George Gamow had solved many imposing scientific problems before he turned his attention to cracking the code of life. Born in 1904 in the Black Sea port of Odesa, Gamow began contemplating the universe at age six when he saw Halley's Comet from the rooftop of his family's apartment building. Four decades later, he would become the world's leading proponent of the theory that the universe began with a "Big Bang." Gamow's fellow scientists regarded him as a genius—"another Heisenberg," said Niels Bohr, comparing him to the Nobel Prize–winning pioneer of quantum mechanics; but also an oddball, "a giant imp, jumping from atoms to genes to space travel," said Jim Watson.

Gamow stood out for his 6'6" height and for his puckish sense of humor, displayed even in the most serious academic settings. When publishing the theory of the cosmological origin of the chemical elements he had developed with his student Ralph Alpher, Gamow popped in the name of his colleague Hans Bethe, simply to create an Alpher-Bethe-Gamow author list befitting the Greek alphabet.

After defecting from the Soviet Union in 1933, Gamow arrived in the United States a year later. A professor of physics for 20 years at George Washington University in Washington, D.C., Gamow eventually made his way to my own University of Colorado in Boulder. Though he had cut his teeth on nuclear physics and cosmology, Gamow became convinced by the early 1950s that the most exhilarating scientific question left to answer had nothing to do with the origin of the universe or the behavior of subatomic particles. In fact, it had nothing to do with physics at all.

In June 1953, Gamow read Jim Watson and Francis Crick's momentous article in the journal *Nature* announcing that the structure of DNA was a double helix. (This breakthrough owed much to the British crystallographer Rosalind Franklin, whose role in the discovery has only recently been broadly recognized.) The structure provided a neat solution to a major mystery: how genetic information is duplicated so it can be passed from one generation to the next. The four chemical units, or *bases*,* arranged along a DNA strand—adenine (A), thymine (T), guanine (G), and cytosine (C)—form *base pairs* with *complementary* bases on the other strand of the double helix: A always matches with T, and G with C.

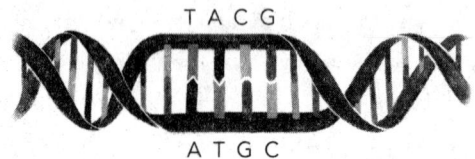

Base pairs hold together the two strands of the DNA double helix. The section in the middle is unwound to highlight the A-T and G-C base pairs.

The double helix looks like a twisted zipper. If one were to unzip the zipper, each side would have all the information needed to direct the con-

---

*The same abbreviations—A, T, G, and C—are used for both the bases and the nucleotides: a nucleotide is a base attached to a deoxyribose sugar and a phosphate group. The full chemical names of the nucleotides are adenosine, thymidine, guanosine, and cytidine. Although the distinction between nucleotides and bases has biochemical importance, their information content is identical.

struction of the opposite side, as the two sides are always perfectly complementary. This must be how genetic information is replicated, Watson and Crick reasoned.

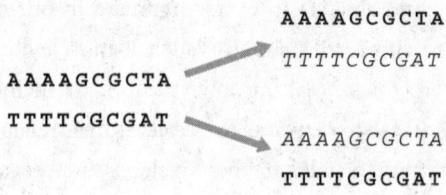

Boldface letters represent parental DNA, and italic letters represent newly synthesized daughter DNA.

In a letter dated July 8, 1953, Gamow congratulated Watson and Crick for bringing biology into the "exact sciences." He boldly proposed a collaboration across disciplines to tackle the next big question that was on everyone's mind: how the information encoded in those strings of A, T, G, and C was actually read out to specify ultimately a hand, a heart, a liver, or a brain—or the wrinkled and smooth peas in the garden of Gregor Mendel, the Augustinian monk who first proposed that such traits were passed down from generation to generation in the form of some elementary unit, which we now call a gene. Gamow suggested that he could help Watson and Crick use math and physics to decrypt this genetic code.

Watson and Crick were flattered that a famous physicist was interested in their work, but Gamow's handwritten letter "had so many whimsical qualities," Watson recalled, "that we did not know how serious he might be." Gamow was very serious. Over the next several months, he threw himself into deciphering the genetic code. He was a consultant to the U.S. Navy at the time, so he sought out not only chemists and physicists to help in his quest but also military cryptographers. As it would turn out, however, solving this secret of DNA would ultimately require DNA's own daughter—ribonucleic acid, or RNA.

A RIDDLE, WRAPPED IN A NECKTIE

In biological terms, "deciphering the genetic code" means understanding how DNA encodes *proteins*. Life is built from proteins, which are the main movers and shakers in every organism in our biosphere. In humans, some proteins form structures such as muscle fibers, skin, and hair. Some act as enzymes, breaking down the food we eat into its constituent components and then recycling these pieces to build up new cellular machines. Others punch holes in the envelopes that encase our cells, selectively allowing some salts or nutrients to enter the cell and expelling others. Other proteins act as signaling molecules, receiving information from the outside world and activating internal processes accordingly. Still others are antibodies, which protect us from foreign invaders such as viruses. In short, proteins are spectacularly diverse.

Chemically speaking, proteins are polymers, strings of a hundred or even a thousand amino acids. These amino acid building blocks come in 20 types, with names such as lysine, valine, and phenylalanine. Each protein has a specific order, or sequence, of amino acids along its chain. This sequence determines how the protein folds into a three-dimensional entity that has a particular function, such as digesting food in the stomach or transmitting neuronal signals in the brain. Thus, even though we hear that we need to eat fish or tofu or an Impossible Burger "to get enough protein in our diet," there is not *a* protein but rather many thousands of different ones. And each is encoded by its own gene, which is made of DNA.

Whereas proteins are built out of 20 amino acids, DNA is built out of the four bases. This leaves us with, on the one hand, the sequence of A's, G's, T's, and C's in a gene made of DNA and, on the other hand, the sequence of 20 amino acids in the corresponding protein. Gamow sought to figure out how different arrangements of the four DNA bases specified, or encoded, each of the 20 amino acids. By 1954, he had recruited Watson, Crick, and an eclectic cadre of well-known scientists—20 in all, one for each amino acid—to direct their brain power to the coding problem.

Gamow had woolen ties, each embroidered with a strand of RNA, sewn for the members of this little fraternity of pointy-headed code breakers. He dubbed the group "the RNA Tie Club." Why not the "DNA Tie Club," you may ask? It was Jim Watson who convinced Gamow that they had to focus their code-breaking attention on RNA because of the basic structure of cells. Whereas DNA is found *inside* the cell's nucleus in higher organisms, proteins are made *outside* the nucleus, in the region known as the cytoplasm. This spatial separation mandated that some sort of messenger must transport the information from DNA to the site of protein production. A number of forward-thinking scientists thought that RNA could be that messenger, as it was known to be abundant both in the nucleus and in the cytoplasm.

Since the early 1900s, chemists had analyzed RNA and DNA and found that they were cut from the same biochemical cloth. Their full names—*deoxy*ribonucleic acid and ribonucleic acid—reveal how closely related they are. "Deoxy" indicates that DNA has one fewer oxygen atom on each of its repeating units than RNA; the extra oxygen atom makes RNA chemically much less stable than DNA.

For many decades, DNA and RNA were seen as chemical compounds without a function. They were far less interesting to scientists than proteins, whose roles as enzymes and as signaling molecules such as insulin seemed so exciting. DNA emerged as the centerpiece of biology only when the 1944 discovery by Oswald Avery and colleagues at the Rockefeller Institute that DNA is the molecule responsible for causing heritable changes in bacteria was coupled with the 1953 discovery by Watson and Crick of the double-helical structure of DNA. Because scientists had already theorized as early as 1947 that RNA was copied from DNA, they now assumed that RNA must also have an important role to play in the chemistry of life.

Although RNA is typically single stranded and DNA is double helical, both speak the same language. Like DNA, RNA is written in a four-letter alphabet. The first three letters—A, G, and C—are the same as DNA's. The fourth letter, U (uracil), occupies the same place in RNA's alphabet as T does in DNA's alphabet. So when RNA is copied from DNA, it carries the same information.

Devoted as always to big, sweeping theories, Gamow first tried to work out a solution to the coding problem with nothing more than pencil, paper, and his brain. In 1953, the same year the double helix was discovered, he theorized that three bases might code for one amino acid. His reasoning was mathematical. If you took the four letters in the DNA or RNA alphabet and tried to create as many combinations of two letters as possible, you would get only 16 combinations—too few to specify the 20 amino acids that make up proteins. But if you tried to arrange the four letters of the DNA or RNA alphabet into as many three-letter combinations as possible, you would get 64 of these *triplet codons*—enough to specify 20 amino acids and then some. Longer groups of bases could do the job, but three were enough; it was the most economical format for the genetic code.

But which groups of three bases encoded which amino acids? Even though Gamow had brought together some of the greatest geniuses of the twentieth century, ultimately the distinguished members of the RNA Tie Club succeeded only in tying themselves in knots. By the late 1950s, Gamow had given up trying to crack the code, convinced that the problem had "no solution" that could be found on "the basis of pure theory."

What Gamow really needed was the biological equivalent of the Rosetta Stone. Instead of a text written in both Egyptian hieroglyphics and Greek, it would display an *amino acid sequence* of a protein as well as the *RNA sequence* that encoded it. But there was no such stone; scientists would need to devise their own. And to do it, they would have to determine whether this hypothetical "messenger" actually existed—whether RNA truly was the missing link between our genes and the protein building blocks of life.

DID YOU GET MY MESSAGE?

In the 1950s, plenty of scientists, including Jim Watson and Francis Crick, fervently believed in the theory that RNA might carry information from the DNA in the cell's nucleus to the cytoplasm, where proteins were synthesized. But when scientists tried to find empirical evidence that RNA is the messenger of life, their initial results fell flat. Their first disappointment came when they found that most of the RNA in a cell's cytoplasm has the same proportion of A, G, C, and U bases independent of what proteins were being synthesized. This didn't make sense—it would be akin to finding that Beethoven's Ninth Symphony had exactly the same proportion of each kind of musical note as Lady Gaga's "Bad Romance." You would expect that these very different pieces of music would have correspondingly dissimilar distributions of F-sharps and B-flats and so on. Similarly, you would expect that different proteins, with their different amino acid compositions, would have different proportions of A, G, C, and U in the messenger RNAs, or mRNAs, that specified them.

The second disappointment was that most of the RNA in the cytoplasm is very stable—once born, it has great longevity. But scientists had seen cases where the proteins being made in a cell switched rapidly, within a few minutes, from one set of proteins to a completely different set. For example, if you switched the food source for bacteria, they would stop making the enzymes they used to digest the old food and immediately start making enzymes suited to the new food. Likewise, if a bacterium became infected with a virus (bacteria are plagued by their own viruses, called *bacteriophage*, or *phage*), it would switch from producing bacterial proteins to producing phage proteins. Consequently, a true mRNA would need to be unstable to allow rapid changes in what proteins were being made. The rock-stable nature of most of the cellular RNA seemed to disqualify it from being the messenger the scientists sought.

Among the scientists who remained convinced that there must be an mRNA hiding from view were François Jacob and Jacques Monod of the Pasteur Institute in Paris. They had made seminal discoveries of how genes were turned on and off in bacteria and were now turning their attention to RNA. In 1960, on a visit to Cambridge, England, Jacob met with his friends Francis Crick, Sydney Brenner, and others in Brenner's rooms at King's College. Jacob described his latest experiments on how genes were regulated in bacteria, and the conversation soon turned to excited speculation about the role of messenger RNA as the link between gene and protein.

Suddenly, and almost simultaneously, Crick and Brenner jumped up. They recalled recent experiments by two scientists at Oak Ridge National Laboratory in Tennessee, Ken Volkin and Larry Astrachan. In studies of *Escherichia coli* infected by a phage called T2, Volkin and Astrachan had witnessed the rapid formation of new RNAs that were smaller than the stable RNA in the cell (now known to be the RNA embedded in *ribosomes*, the cell's protein factories). For a host of complicated reasons, Volkin and Astrachan had interpreted their data as indicating that RNA was metamorphosing into DNA. But what if, instead, they had caught a glimpse of the elusive mRNA?

It was an exciting possibility, but a more direct test of the mRNA hypothesis would be necessary to confirm it. Sydney Brenner and François Jacob were set to visit the geneticist Matt Meselson at Caltech a few weeks later—it was decided that the three of them would perform the critical experiments. Meselson and his colleague Frank Stahl had recently used a new technique, involving an extremely fast-spinning centrifuge, to test Watson and Crick's conjecture that the strands of the double helix came apart during DNA *replication*. Their results supported the theory, finding that each progeny double helix retained one "old" strand from its parent and paired with one newly synthesized strand.

Now, Brenner, Jacob, and Meselson would use this same ultracentrifuge to try to find the mRNA needles in the ribosomal RNA haystack. The plan was to infect *E. coli* bacteria with a phage, which would cause the bacteria to switch the type of protein they were producing. At the same time that the scientists added the phage, they would add a radioactive version of uracil (labeled with carbon-14) that would be incorporated only into newly made phage mRNA, not into preexisting bacterial RNAs. If the mRNA hypothesis was correct, then new phage mRNA would appear at just this moment—allowing them to capture this missing link between DNA and protein.

Indeed, Brenner, Jacob, and Meselson saw the radioactive phage mRNA in the ultracentrifuge, clearly smaller than the RNA of the ribosomes. As predicted, this RNA was short-lived, and it teamed up with the preexisting ribosomes of the infected bacteria to produce the new kinds of proteins needed by the phage to do its dirty work. They had finally captured mRNA, as elusive as it was.

One way of thinking about this biological process is to imagine a record player. The ribosome is your turntable, the mRNA is the vinyl LP record, and the protein is the music you hear when you lower the needle. You might change the record depending on your mood, but the rest of the setup remains the same. Just as a record player can play any LP record, ribosomes can work with any mRNA that comes along. The mRNA is what determines the specific protein produced—whether a phage protein, an *E. coli* protein, or something else—just as the record determines what music you're hearing.

The reason mRNA was so difficult for scientists to detect is that only about 5 percent of the RNA in an *E. coli* bacterium is messenger RNA. The other 95 percent is dominated by ribosomal RNA. Moreover, *E. coli* has 4,000 different genes, each of which yields an mRNA of a different size and sequence. Thus, any given mRNA is far less than the 5 percent total. Finally, most *E. coli* mRNAs have a life span of merely a few minutes, making them elusive to capture. In contrast, ribosomal RNAs are not only omnipresent in the cytoplasm but also very long-lived—so it's easy to see why it took so long for pioneering scientists to see past all that ribosomal RNA.

A LINGUISTIC INTERLUDE

Thinking about a record player might help us explain how ribosomes and mRNA team up to make proteins. But when it comes to understanding how RNA encodes proteins, we need to switch off the music and crack open a book.

The pages of most books are filled with written language—letters, words, sentences. In the book of life, which is written in the language of DNA, we don't have an alphabet of 26 letters as in English or 22 letters as in Hebrew but four letters: A, G, C, and T. The way those four letters are arranged—their sequence—will determine the *meaning* of each word and each sentence.

So how many books do we need to capture the meaning of life—to capture *all* the DNA in a given organism? The answer depends, of course, on the size of the *genome*, a genome being the totality of the DNA in an organism. The human genome, bundled into 23 chromosomes,* has about 3 billion bases, or letters.

Assuming a typical font size that allows about 3,000 characters per page, it would take 1 million pages to record the human genome. Now, considering that a long book might have 500 pages, the genome sequence would require about 2,000 books—too many for your home library, but just a fraction of what your local public library keeps on hand. Each human chromosome, comprising a single linear DNA molecule, would occupy about 90 books. A bacterial genome, such as that of *E. coli*, is a lot smaller; the whole genome resides in a single circular chromosome. Containing 4.5 million bases, it would fit comfortably into three library books.

As we've learned, to make a protein we need messenger RNA to take information written in DNA to the ribosome. Naturally occurring RNAs are not copies of entire pages in the grand book of DNA. Rather they have specific start and end points that are unlikely to coincide with page breaks. Thus, rather than a photocopy of a page from a physical book, we might think of mRNA as a section of text copied from an e-book into another electronic document by just a few clicks. We move our cursor to highlight a particular portion of a page and then paste it onto a fresh page in our word processor. (And with the handy-dandy find-and-

---

*Chromosomes consist of DNA molecules packaged up with proteins. Human sperm and egg cells have one set of 23 chromosomes, while body cells have two sets—one from Mom and one from Dad.

replace function, we can simultaneously change all the T's in DNA to U's in RNA, mirroring the chemical process that occurs in nature.)

This copy-and-paste process—in which DNA is being copied into mRNA—is happening constantly in our bodies, every time a new protein needs to be synthesized. The region of the DNA that is copied into mRNA differs depending on what's needed at a particular time and place. Different parts of the genome are copied in growing children compared with the parts copied in full-grown adults, and different parts of the genome are copied in the heart compared with the parts copied in the brain, the liver, and the skin. In other words, the copying is highly regulated.

With our copy-and-paste done, we can examine the sequence of our mRNA. It's an array of A, G, C, and U, such as

```
GUAGGGCAUGCCUUCGAAAAUAUUUUGUUAGCGCCUCCUUGGAGUAGAA
```

Let's say we know how to decode the groups of three bases—the triplet codons—along this mRNA. Here's our dictionary. Instead of amino acids, I've chosen three-letter English words as the meaning of each triplet codon:

```
AUG = The
CCU = big
UCG and UCC = cat
AAA and AGA = ate
AUA = one
UUU = fat
UGU and UAU = rat
CGC = but
CUC = two
CUU = and
GGA = six
GUA and GUU = for
GAA and GAG = you
```

```
AAU = now
AUU = see
UUG = fox
UUA and UAA = run
GCG and GCC = out
UGG = fun
AGU and AGC = sun
```

My dictionary has an exceedingly limited vocabulary—only 20 words, analogous to the 20 natural amino acids. Some of these words are encoded by a single codon: *fun* is encoded by UGG. Other words (*run, out, sun,* and so on) are encoded by more than one codon, just as in the genetic code. You might ask why nature would evolve a system that uses so many codons—64 in total—to encode only 20 words (or amino acids). Isn't there a more graceful way of structuring the code?

This seeming inelegance is one of the things about biology that tripped up physicists such as Gamow. In physics, events are largely predictable. If you know Maxwell's equations, you can solve your classical physics problem. But with biology, the only rule is: whatever works. Once a system, however convoluted it may seem, starts to work well, it gets locked into place by evolution and becomes very hard to change. The fact that a good engineer could devise a more straightforward or more efficient system is irrelevant.

Now that we know how to translate the message from strings of codons into meaningful words, we need to know where the message starts and where it ends. Do we start at the left end, reading GUA as "For," and continue reading to the right end? While that might seem reasonable, nature has a different solution. It uses the triplet AUG to designate the start of a sentence. Our equivalent rule here is that sentences must start with "The," which is encoded by AUG. We then simply scan from the left end, look for the first AUG and start reading there, with the dictionary in hand.

```
GUAGGGC AUG CCU UCG AAA AUA UUU UGU UAG CGC CUC CUU GGA GUA GAA
        The big cat ate one fat rat
```

Everything's fine until we reach the UAG codon and find no entry for it in our dictionary. Oh, yes, we need another triplet codon to mark the end of the sentence—a period.* Let's say

```
UAG = end of sentence = period
```

Our decoding is now complete:

```
GUAGGGC AUG CCU UCG AAA AUA UUU UGU UAG CGC CUC CUU GGA GUA GAA
        The big cat ate one fat rat.
```

Now that we've specified start and stop codons—AUG and UAG—we see that separating the codons by spaces is unnecessary. If we are given the mRNA sequence as an uninterrupted string of letters, and know that we read it three letters at a time, we still get the same information:

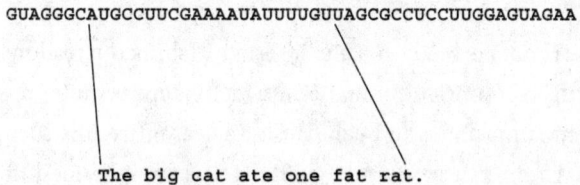

```
GUAGGGCAUGCCUUCGAAAAUAUUUUGUUAGCGCCUCCUUGGAGUAGAA




        The big cat ate one fat rat.
```

Even the triplet nature of the code that Gamow hypothesized remained only a good guess during the 1950s. In 1961, a clever set of experiments by Francis Crick and his colleagues involving the mutation of a phage gene by acridine dyes finally demonstrated the triplets. Acridine dyes are flat molecules much like DNA bases, so they slip into the

---

*Note that the nucleotides preceding the start codon and following the stop codon may not be parts of other "sentences" (or genes), but may have regulatory functions.

double helix and can be mistaken for bases when the DNA is replicated—thereby causing a wrong base to be inserted in the daughter DNA. Consider a row of books packed next to each other on a bookshelf. If you squeeze in another book, all the ones that follow it must move over one position to accommodate the insertion. If you squeeze in two, the others shift over by the width of two books.

Our linguistic analogy can show us how such insertions would affect the message and how multiple insertions support the idea of a triplet code. Starting with our hypothetical message, I've marked the start and stop codons with italics:

```
GUAGGGC AUG CCU UCG AAA AUA UUU UGU UAG CGC CUC CUU GGA GUA GAA
        The big cat ate one fat rat.
```

Here's what happens when the acridine dye causes a U, underlined below, to be randomly inserted:

```
GUAGGGC AUG CCU UCG UAA AAU AUU UUG UUA GCG CCU CCU UGG AGU AGA
        The big cat run now see fox run out big big fun sun ate
```

Our sentence starts out correctly, but the inserted U shifts the reading frame. We call this a "frameshift mutation," which happens quite frequently in the human genome and can cause unpleasant results such as cystic fibrosis, Crohn's disease, and Tay-Sachs disease. A frameshift is so deleterious because it renders all the words after it nonsense. Furthermore, the UAG that previously designated the end of the sentence is now out of frame, so the nonsense sentence goes on and on. If this were mRNA encoding a protein, the protein would be useless.

Now let's see what happens when we add *two* acridines to our sequence, causing *two* bases to be inserted. This is just as destructive as a single-base insertion:

```
GUAGGGC AUG CCU UCG UUA AAA UAU UUU GUU AGC GCC UCC UUG GAG UAG
        The big cat run ate rat fat for sun out cat fox you.
```

In this case, the frameshift produces a UAG stop codon at the right-hand end, but the resulting sentence is still nonsense. If this were mRNA, the resultant mutated protein would once again be useless.

Finally, let's examine the consequences of a three-base insertion:

```
GUAGGGC AUG CCU UCG UUU AAA AUA UUU UGU UAG CGC CUC CUU GGA GUA
        The big cat fat ate one fat rat.
```

If the code is a triplet code, as shown, one aberrant word—"fat"—gets inserted into the sentence, but all the other words are correct. The sentence is more or less understandable.

And that's exactly what Crick and his colleagues reported, confirming Gamow's original hypothesis. One insertion was deadly. Two insertions were also deadly. But three insertions were more or less functional! Thus, the codons must come in groups of three bases.

CRACKING THE CODE

Gamow and his RNA Tie Club had figured out that the genetic code was written in three letters and that RNA was the likely bridge between DNA and proteins. Brenner, Jacob, and their colleagues had proved the latter—conclusively identifying messenger RNA, which feeds the information in our DNA into the cell's protein-building machinery, the ribosomes. But we still couldn't read that information. As the 1960s dawned, the genetic code remained indecipherable—until, that is, a young scientist named Marshall Nirenberg came along.

In 1959, Nirenberg landed an independent research position at the National Institute of Arthritis and Metabolic Diseases in Bethesda, Maryland. Before he turned to biochemistry and earned his doctorate, he had obtained a master's degree at the University of Florida studying caddis flies, a group of insects with aquatic larvae and mothlike adults. (Perhaps Nirenberg was reluctant to acknowledge that he had begun his scientific career working on bugs, as he omitted this fact from his eventual Nobel Lecture.) By trying to crack the genetic code, Nirenberg was making an ambitious move. He had little connection to the leaders of the field and had never received a woolen RNA tie. Nevertheless, departing from Gamow's pen-and-paper approach, Nirenberg decided to tackle the coding problem using biochemistry. This would require re-creating protein synthesis—the process by which mRNA is turned into protein—in test-tube experiments.

Nirenberg built on the groundbreaking work of fellow biochemists Elizabeth Keller and Paul Zamecnik at Massachusetts General Hospital in Boston, who had developed a method for synthesizing proteins outside of cells using ingredients obtained from rat livers. Zamecnik and others subsequently showed that extracts of *E. coli* could be used to do the same. This became Nirenberg's system. In short, the approach involved smashing open *E. coli* bacteria to obtain a crude preparation of ribosomes, then adding different RNA sequences as templates for protein synthesis.*

After multiple false leads, Nirenberg and his postdoctoral fellow, Heinrich Matthaei, stumbled onto the fact that a simple RNA molecule composed entirely of U bases, known as poly(U), directed the synthesis of a single protein composed entirely of the amino acid phenylalanine. The beauty of poly(U) was that no matter where reading started, there could be only one kind of triplet codon, UUU—and it was translated only into phenylalanine. Thus the first piece of the genetic-code puzzle was cracked: UUU encodes phenylalanine.

Now the path forward became clear: feed different RNA sequences into ribosomes and see what proteins come out. Using this approach, poly(C) and poly(A) were found to encode polyproline and polylysine,

---

*The conditions of his test-tube experiments were such that the ribosomes did not need a start signal on the mRNA (the AUG in our above analogy) but would start randomly along the mRNA chain.

respectively. So the table of codon assignments grew: 3 down, 61 to go. Of course, these first three were easy, because each letter in the triplet codon was identical. The more complicated codons required a different approach.

Enter Gobind Khorana.* Born into a poor Hindu family in Raipur, which was then in India but now part of Pakistan, Khorana had been educated in England and Switzerland. He first rose to prominence for his work on cracking the code while a professor at the University of Wisconsin–Madison. He and his group had invented methods for chemically synthesizing DNA, piecing together the *nucleotides* one at a time. They then used a recently discovered enzyme to copy the DNA into RNA. The beauty of this method was that it allowed them to precisely dictate the base sequence of the DNA and therefore of the RNA copied from it. These RNAs, put into the test-tube protein-synthesis system, then acted as messengers to specify the corresponding amino acid sequences. By synthesizing every possible permutation of three bases, feeding them into the test-tube protein-synthesis system, and looking at what string of amino acids came out, Khorana helped fill in the table of the genetic code—thereby earning a share of the 1968 Nobel Prize in Physiology or Medicine alongside Marshall Nirenberg.

After reading about these discoveries in the news, Gamow was satisfied that someone had finally solved the coding problem, even if he couldn't help taking a swipe at the experimental approach. "The solution looks considerably less elegant than the simple theoretical correlation which I had originally visualized," he said. "But it has the indisputable advantage of being correct."

---

*I first met Khorana at MIT when I was a biology postdoctoral fellow in building 16 and he was a Nobel Prize–winning chemist in building 18. As a nod to the severity of Boston winters, many of the MIT buildings are interconnected, so I could walk to the chemistry department without bundling up. I found Khorana to be modest, infectiously excited about his research, and not at all put out that a lowly postdoc who grew up in Iowa would introduce himself.

A LITTLE RNA WITH AN IMPORTANT JOB

Elucidating the code that mRNA uses to specify proteins was a huge accomplishment, but it was not the only big question that needed answering. Having a code was critical, but then what? There must be a process by which the code was "read out." How did the mRNA get the amino acids into the right position to be strung together to build a protein?

Francis Crick came up with an answer to this question in 1955, and his proposal stands out as a singular triumph of theory in biology. He conjured up a class of molecules that had never been seen solely on the grounds that he thought they must exist as the missing piece in the grand theory of the genetic code. He envisioned a group of "adaptor molecules," each with two ends. One end would be linked to one of the 20 amino acids. The other end would recognize and bind to the matching ("cognate") mRNA triplet codon.

You can think of these amino-acid-to-mRNA adaptors a bit like the adaptors that connect electronic devices to power outlets. To charge a pair of headphones, you need the right adaptor for the right plug, and that plug may be different depending on whether you're in the United States or the United Kingdom or Germany. As with headphones, so too with amino acids. You have an outlet (a triplet codon) and a three-pronged plug (called an *anticodon*) that fits inside it. Because the codon and anticodon are complementary, they connect to each other by base-pairing. And like a power strip accommodating multiple devices, the adaptor molecules line up to connect mRNA codons to their cognate amino acids in growing chains that eventually form proteins. Or at least that's what Crick proposed.

But theory can only take us so far. Someone needed to go into the laboratory and see if such an adaptor could be found. That someone was Paul Zamecnik, who had developed the protein-synthesis method that powered Nirenberg's discovery of the first codon. He started with amino acids that had been labeled with a radioactive isotope of carbon, so that he could track where they went—a bit like a bank hiding an exploding

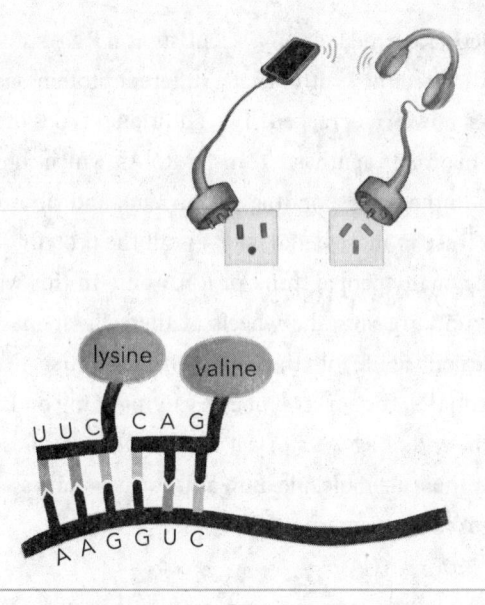

Francis Crick proposed that each triplet codon on an mRNA strand (bottom) would need to be recognized by an "adaptor molecule" carrying the corresponding amino acid, such as lysine or valine, shown here. This is analogous to charging cables for cell phones or headphones that serve as adaptors, connecting various three-pronged electrical outlets with the devices.

canister of paint in a moneybag so the money can be tracked if it's stolen by thieves. When Zamecnik added these radioactive amino acids to his rat-liver system, an amazing thing happened—small RNAs became radioactive, indicating that an amino acid had been joined to an RNA. Such a linkage had never been observed before, but it was exactly what would happen if there was an RNA whose job it was to adapt RNA codons to amino acids. Later these small RNAs became known as *transfer RNAs* (tRNAs), because they transfer the correct amino acid to the growing protein chain inside the ribosome. As we'll see again and again, transfer RNAs may be small, but they are mighty.

By the mid-1960s, it seemed to some scientists as if RNA's story was over. Each mRNA directed the synthesis of a different protein on the basis of a code that had now been cracked. Two additional types of stable RNA were involved in protein synthesis: transfer RNAs, which connected each mRNA codon to the corresponding amino acid, and ribosomal RNAs, which—as we'll see in more detail later—built the protein.

Even today, many people think of RNA only in this way. They see it as DNA's grunt, greasing the wheels of the cellular machinery that transforms the code of life into the stuff of life. Of course, that biological process is essential to the survival of every living thing on Earth. And if it truly were the end of RNA's story, it would still be a colossal achievement for our minuscule molecule. But, as it turns out, messaging is only the first of RNA's many superpowers.

Chapter 2

SPLICE OF LIFE

It is hard to imagine a more idyllic backdrop for a scientific symposium than the Cold Spring Harbor Laboratory. Perched on Long Island Sound, the place looks more like a summer camp than a research center—shingled cabins, sailboats lolling in the breeze, and rolling greenswards that light up in spring with blooming magnolias, azaleas, and dogwoods. Founded in 1890, the laboratory always had a reputation for serious, if somewhat sedate, research. This was the place where, beginning in the 1940s, the reclusive geneticist Barbara McClintock carried out her painstaking experiments to track variations in the color of corn kernels, which she used to identify how genes and chromosomes functioned within cells. Her discovery of "jumping genes" eventually led to a Nobel Prize.

The somnolence of Cold Spring Harbor Laboratory changed dramatically when Jim Watson became director in 1968. Watson was so insightful, so ambitious, and so pushy that he quickly turned the lab into a powerhouse that churned out one major scientific breakthrough after another. At the same time, the lab became a renowned conference center where molecular biologists would share their latest discoveries. It was and still is the place to hear what is new, without having to wait for it to be published many months later, and to forge fresh collaborations.

The topic of the June 1977 Cold Spring Harbor Symposium was the structure and function of chromosomes in higher organisms. A postdoctoral researcher at the Massachusetts Institute of Technology studying this very subject at the time, I was thrilled to be able to attend the symposium. I still remember the excitement when two teams of researchers—one led by my neighbor at MIT, Phil Sharp, and another by Cold Spring Harbor's own Rich Roberts working with Louise Chow, Rich Gelinas, and Tom Broker—took the stage and revealed the secret to a mystery that had haunted scientists for more than a decade.

What was this big mystery? Once the relationship between DNA, resulting mRNA, and protein had been worked out in *E. coli* and its phages, many scientists had moved on to study more complex organisms, such as plants, animals, and especially humans. They had every reason to expect that the fundamental features of biological information storage and transfer would be conserved in all life-forms. As the Nobel laureate Jacques Monod once put it, "Anything found to be true of *E. coli* must also be true for elephants." But when biochemists studied human cells that were grown on Petri dishes in incubators, they became confused. They expected that mRNA would first appear in the cell nucleus, where its chromosomal DNA parent was located. Indeed, they found RNA there, but it seemed too large to be mRNA—it was on average 10 times bigger than what was required to code for a protein. This was strange, because the mRNA that had been exported from the nucleus to the cytoplasm was just the right size to make a protein.

Was this oversized RNA found in the cell nucleus really going to turn into mRNA? If so, what were all the extra nucleotides in the nuclear RNA doing if not coding for a protein? As revealed that day at Cold Spring Harbor, the answer would provide the first hint that RNA was capable of a lot more than simply delivering messages at DNA's service.

UNDER WRAPS

Though I happened to work across the street from Phil Sharp's lab at MIT, this didn't mean I had a front-row seat to Phil's discovery. Occasionally, I would wander across Ames Street to his lab in the Cancer Center to ask for advice about troubleshooting experiments. My friend Claire Moore worked there, and she and Phil were always happy to talk about my research—but they were unusually quiet about their own results.

This was odd. Scientists usually can't help but blab about their research, the exciting discovery that just might be around the next corner. But Phil, Claire, and postdoctoral fellow Sue Berget knew they were onto something big, so they had decided to keep quiet. It was only by sitting in the auditorium at Cold Spring Harbor a year later that I would find out how Phil and the group of Cold Spring scientists solved the riddle of human mRNA's mismatched size.

The key turned out to come from adenovirus, a DNA-based virus that gives humans the common cold. Just as bacteriophage had given the first molecular biologists a handle on how bacteria went about their genetic business, human viruses provided a gateway into studying the molecular details of human biology. After all, phages and human viruses both work by tricking their host cells into providing the machinery that powers their infectious cycle, so these parasites must use the same fundamental biological mechanisms as their hosts. Study of viruses offers practical advantages, too: infected cells contain many copies of the viral DNA and corresponding RNA products, giving researchers a lot of material to work with.

Both the MIT and the Cold Spring Harbor groups started by mapping where various genes were located on the adenovirus chromosome. They didn't expect this to lead to any big discovery; the mapping was meant only to provide the necessary framework for later work to understand how the viral genes were expressed. When the researchers compared the viral DNA to its mRNA copy found in the cell cytoplasm, they expected to find the DNA and mRNA sequences marching along in sync from one end to the other. Certainly, that was what would have happened in bacteria.

Instead, they discovered that large internal portions of the mRNA—which would have been present if the mRNA were a straightforward copy of the DNA—had simply vanished. It appeared that they'd been "spliced," with some middle parts cut out and the flanking sequences joined together. The researchers were forced to conclude that the coding regions of the adenovirus gene were not continuous but rather were split up and separated by stretches of noncoding DNA that we now call *introns*.

The audience at Cold Spring Harbor Laboratory was stunned. Jim Watson himself, who was in the room that day, called the discovery a "bombshell." Messenger RNA was *supposed* to be a continuous copy of its gene. It seemed incredibly inefficient for it to be anything else, and it was difficult to conceive why protein-coding regions of DNA would be split by introns or how the introns would be spliced out. Were all these

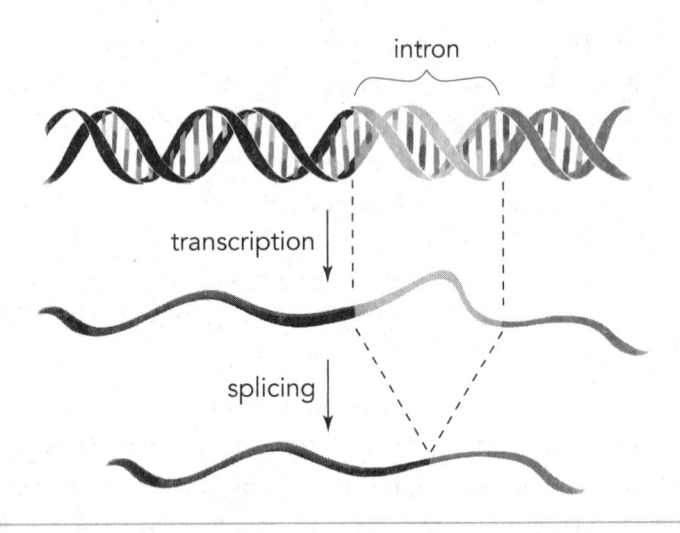

Stretches of noncoding DNA called *introns* interrupt most protein-coding genes in humans and many other eukaryotes. Each intron (light shading) is transcribed as part of the precursor RNA (middle), which is then spliced to give the final functional mRNA (bottom).

complicated operations, in which bits of code were shuffled into DNA and out of mRNA, just idle acrobatics—some evolutionary dance leading to nowhere? Or did this splicing process perhaps serve some greater purpose? For a time, the world of molecular biology was turned upside down trying to answer these questions.

Soon the stakes got even higher. Scientists began to recognize that introns were not limited to viruses but were a common feature in *eukaryotes*—that is, organisms that form a cell nucleus to sequester their DNA. Once the existence of adenovirus introns was announced, many scientists in labs around the world realized that the genes they had been studying were similarly split by introns. For example, later in 1977, National Institutes of Health biologists Shirley Tilghman* and Phil Leder found that a mouse gene encoding the blood protein hemoglobin had its coding region split by two introns. Both the introns and the coding sequences were copied from the gene to form a long RNA in the cell nucleus—but then, before the mRNA was exported to the ribosomes to make proteins, Mother Nature took out her pruning shears and magically spliced out the interruptions.

Phil Sharp used the word *splicing* to describe this process to convey the same sort of repair job that a sailor might perform on a badly frayed segment of a rope. The sailor might cut the rope above and below the frayed segment, throw away the damaged piece, and attach, or splice, the ends of the two good pieces back together.

Alternatively, we might think of an intron as a few meaningless words, a string of "blahs," interrupting an otherwise intelligible sentence: *You really smell nice blah-blah-blah-blah-blah today.* With a word-processing system, we can fix this quickly: just highlight the offending interruption, press "Delete," and the *blahs* are spliced out. *You really smell nice today.* Nature uses an analogous process to edit introns out of mRNA, leaving a clean genetic code that can be used to make a protein.

---

*This is the same Shirley Tilghman who, because of her other talents, later became president of Princeton University.

Now scientists understood why the RNA initially made in the nucleus—which contained all those introns—was much larger than the mRNA, which had the introns spliced out. But as usually happens in science, answering one question led to another. What was the function of these introns? Were they worthless, as some scientists assumed? Or might they actually be a key to understanding not only the difference between an elephant and *E. coli* but also what makes us human?

THE PERPLEXING SIZE OF THE HUMAN GENOME

A *genome* is all the DNA in all the chromosomes of an organism. Before the human genome was sequenced* around the year 2000, we had no clue how many genes humans needed to produce all the essential proteins that make us who we are. Yet this fact did not keep us from making guesses, most of which turned out to be astronomically wrong.

Blame yeast. Yes, the same yeast that makes bread dough rise and beer and wine ferment. Yeast is a single-celled organism with no brain, no heart, no arms or legs, no stomach or liver or intestine. No reproductive tissues, either—a yeast cell reproduces by growing a bud, which gets larger and larger until it breaks off and forms a new cell. It seemed fairly obvious, therefore, that a yeast would need vastly fewer genes to go about its business than a human being comprising hundreds of types of cells.

In 1996, the yeast genome became the first of any of the eukaryotes to be sequenced. The yeast genome was found to consist of about 6,000 protein-coding genes.† The Human Genome Project was just taking off at that time, and bets were being taken. How many human genes would there be? Certainly many more than the simple yeast! Famous scientists on the lecture circuit were predicting 100,000 human genes or more.

---

*Sequencing a genome means reading off the order of the A's, G's, T's, and C's on the DNA molecules residing in all of the organism's chromosomes.
†Initially, protein-coding genes are counted by finding stretches of the sequence that are composed of codons. Later, this identification can be confirmed by finding the protein that has the predicted sequence of amino acids.

So you can imagine how shocked scientists were when the human genome sequence was finally announced in 2003, and we learned that humans had about 24,000 protein-coding genes—only about four times as many as the lowly yeast. *Could that be right?* It didn't seem to make any sense. Humans and yeast were both DNA-based organisms. How could we get so much more genetic bang for our buck than our fungal relatives at the bottom of the food chain?

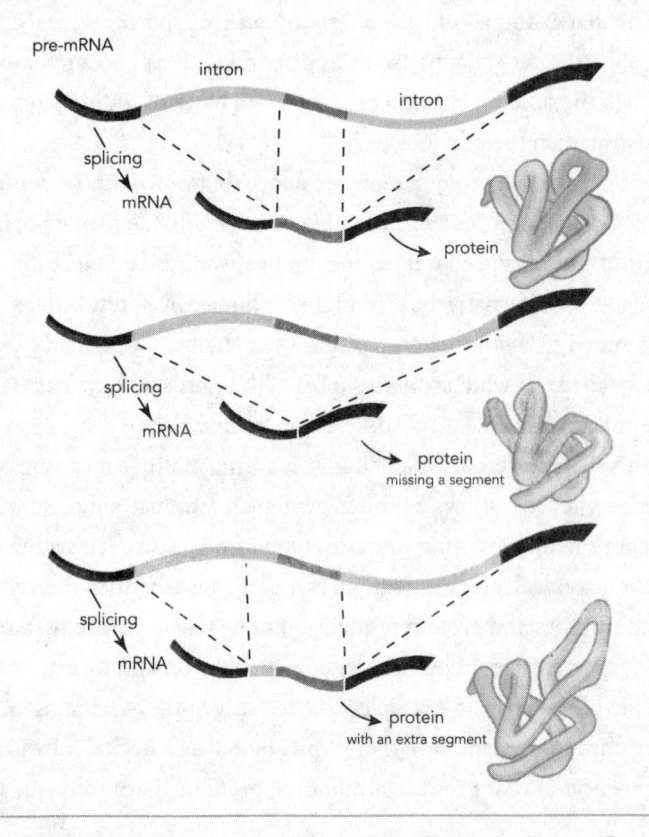

Alternative mRNA splicing gives multiple proteins from a single gene. Because most human genes have two or more introns, RNA splicing can occur in more than one way. Top: Independent splicing of two introns gives an mRNA that encodes one form of a protein. Middle: Splicing skips over one of the coding sequences, resulting in a protein that's missing a central region. Bottom: Use of an alternative splice site results in a longer mRNA and a protein with an expanded section.

A big part of the answer concerns the "nonsense" introns. In yeast, most genes are intronless, and those that have introns typically have just one, so there's only one way for RNA splicing to occur. But in humans, the introns turned out to be more than just an annoyance. They are quite useful. They give RNA wiggle room to splice the code in more than one way, resulting in a wide repertoire of potential proteins from the same set of genes.

The first examples of such *alternative mRNA splicing* were found in 1980, shortly after splicing itself was discovered. It can occur, for example, when the splicing machinery skips over a block of coding sequences, as illustrated on the previous page.

Let's circle back to our sentence above, but now with two introns: *You really smell blah-blah nice blah-blah-blah-blah-blah today.* Normally both introns are spliced out, so the final version reads: *You really smell nice today.* But alternatively, *nice* might be skipped over, resulting in *smell* being joined to *today*. Splicing would then change the sentence to *You really smell today*, which contains most of the words of the previous sentence but conveys an entirely different meaning.

This sort of alternative splicing—obtaining multiple meanings from a single sentence—allows a limited genome to produce a more complex array of proteins, resulting in more complex organisms.* It's a big deal, because a typical human gene gives rise to four or five alternatively spliced mRNAs and protein products. It's one major feature that allows the unexpectedly small human genome to make us who we are.

One of my favorite examples of alternative mRNA splicing occurs in our immune system. *B cells* are white blood cells (technically known as *lymphocytes*) that produce antibodies, proteins that protect us from

---

*It's relatively straightforward to determine which splice sites are used or not used to give rise to alternative splicing, because one simply compares the gene sequence to the mRNA sequences. The challenging question is how alternative splicing is regulated. One model holds that some splice-site sequences are weaker than others. If the amount of a protein factor involved in splicing varies from one cell type to another, this can be enough to cause a weak splice site to be used in one case and skipped in the other.

infections by recognizing and neutralizing foreign pathogens that invade the body. Early in their lifetime, the antibodies are displayed on the outside surface of the B cell, on the lookout for a pathogen—such as a coronavirus—that can bind to that specific antibody. Later, the B cells switch gears and release the antibodies, which circulate through our bloodstream and, again, bind to a complementary shape on a virus. In their first incarnation, the antibodies are standing sentry, while in the second they're in hot pursuit of an invader.

The two types of antibodies—the surface *receptors* and the circulating forms—are identical where they attach to the virus, but they differ at their other end. The type that rides along on the B cell has a greasy end that anchors itself into the cell membrane, while the circulating version has a sugar-coated end that releases it from the cell membrane and allows it to move around in the bloodstream. Both forms are encoded by the same human gene. The RNA copied from that single gene can be spliced in two ways to create two partly identical but distinctly different proteins: *You really smell nice today* versus *You really smell today*. Alternative RNA splicing allows this very efficient dual use of a single gene.

So what's the difference between the elephant and *E. coli*? It all depends on how you splice it.

"U" MARKS THE SPOT

The fact that animals make enormous *precursor* mRNAs only to cut and paste them to make the useful mRNAs was an astonishing discovery. But just because scientists had discovered splicing did not mean they understood how the process worked. A major question was how the introns were identified for elimination. The splicing sites had to be recognized with exquisite precision, because slippage by even a single base would produce a frameshift mutation, changing all the downstream codons and destroying the function of a protein.

Like many scientists who make groundbreaking discoveries, Joan Argetsinger Steitz would use a dose of serendipity to figure out how nature "dropped a pin" at one of the splice sites. Joan had entered the biochemistry PhD program at Harvard in 1963 as the only woman in her class and the first female graduate student in Jim Watson's research group. After completing her PhD on phage RNA, she and her husband, Tom Steitz, who was also a biochemist, went to the famous Laboratory of Molecular Biology in Cambridge, England. There Joan collaborated with the likes of Francis Crick and Sydney Brenner. When the Steitzes then moved on to the University of California, Berkeley, however, it soon became clear that although Tom had a faculty position, Joan would be hired there only as her husband's assistant. As the chairman of biochemistry told her, "All our wives like being research associates." So the two quickly decamped to Yale University in 1970, where they were both given assistant professorships.

At Yale, Joan became interested in the large RNAs made in the cell nuclei of mammals. This was a few years before introns were discovered, so it wasn't yet clear that these large RNAs were precursors to mRNA. Joan wanted antibodies that could be used as tools to identify and potentially inhibit the activity of proteins bound to these large nuclear RNAs. The process involved injecting rats with the protein of interest so that the rat immune system—seeing the human protein as a foreign substance—would make antibodies against it, just as it would do against a viral invader. But her attempts at getting rats to make such antibodies were painful—literally. She still has a scar on one of her fingers where a rat bit her.

Then, in 1979, Joan heard that patients with lupus—a disease in which the human immune system turns against a person's own body—had been found to make antibodies to components present in their own cell nuclei. Could these antibodies be recognizing proteins that bound to nuclear RNAs? She sent a medical student in her lab, Michael Lerner, across the street to the Yale immunology department to fetch serum samples, which would contain antibodies, from such patients.

Joan and her student soon discovered that the lupus patients were making antibodies that targeted proteins bound to small RNAs found in the patients' own cell nuclei. This was bad news for the lupus patients—

we're supposed to be making antibodies against foreign invaders, not against our own cellular constituents—but good news for science, because these so-called *small nuclear RNAs* (or snRNAs for short) would be a gateway to understanding the magic of mRNA splicing. These snRNAs were already known to exist; there were six varieties, and because they were rich in uracil (the letter U in the RNA alphabet), they had been dubbed U1 through U6. But, in 1979, their function was still a mystery.

At this point, it had been two years since the bombshell announcement at Cold Spring Harbor of genes split by introns, and already many additional examples of human introns had been found. Now, everyone wanted to know how cells kept track of which sequences they needed to keep, to make messenger RNA, and which they needed to splice out. This question was a constant topic of discussion in RNA research labs, including Joan's.

Joan knew that introns came in different sizes and sequences. But the introns looked almost identical at their extreme ends, all starting with a sequence similar to GUAAGU and all ending with AG. Joan was keenly aware that single-stranded regions of an RNA have a natural propensity to pair up with single-stranded regions of another RNA using complementary base pairs. It therefore seemed possible that the conserved sequences at the ends of an intron might be recognized and marked as

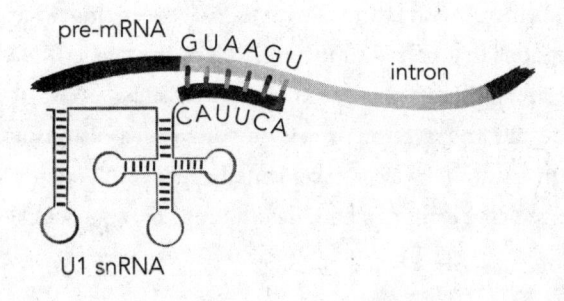

The U1 snRNA locates one end of an intron sequence by base-pairing, thereby marking the spot for mRNA splicing. RNA forms A-U base pairs, which are chemically equivalent to the A-T base pairs of DNA.

splice sites by pairing with one of the U RNAs. So Joan and Michael Lerner scanned the sequences looking for a match, and there it was: the sequence at one end of the U1 snRNA fit together with the sequences at the start of the known human intron sequences, at least on paper. This match seemed too good to be a coincidence, so they made a provocative suggestion: maybe U1 was the agent that identified the beginning of each intron, allowing it to be cut precisely at that spot.

Within a few years, this conjecture held up to a number of tests. For example, Phil Sharp's lab had developed a test-tube system to observe the mRNA splicing reaction, and they teamed up with Joan's lab to see whether the antibodies that recognized the snRNA-protein complexes would shut down splicing. If the U1 snRNA was in fact the entity that recognized the first splice site, the splicing should be inhibited by the antibody. And, indeed, the collaborative effort found just such behavior.

How about the conserved AG at the other end of each intron? The situation there was a bit more complex. The U2 snRNA base-paired with a complementary sequence near the end of the intron, but it was a protein associated with the U2 snRNA that recognized the AG splice site.

The discovery that snRNAs were part of the mRNA splicing machinery added a new function to RNA's repertoire: it can mark sites of interest, akin to dropping a pin on an internet road map. This is a fitting function for single-stranded RNA, because it's always ready to engage in base-pairing, A with U and G with C. We've seen this twice, first with mRNA codons that pair with the anticodons in transfer RNA and now with snRNAs pairing at or near mRNA splice sites.

Yet despite the dazzling precision with which introns are spliced, nothing in nature is perfect. Accidents happen, and when it comes to introns, a splicing error can have devastating consequences.

RNA SPLICING GONE BAD

Typically, if some biochemical process is truly important for a healthy human, there will be diseases that occur when the process fails. Indeed, in 1981—just four years after mRNA splicing was discovered—the blood disorder beta-thalassemia became the first recognized disease that can be caused by mis-splicing.

Beta-thalassemia is one of the most common genetic diseases in Mediterranean countries, the Middle East, and Asia. Patients suffering from the disease are anemic, meaning that they have fewer red blood cells, which leads to low oxygen levels, fatigue, and risk of early death. Hemoglobin, the protein in our red blood cells that carries oxygen, is assembled from four chains of amino acids: two identical *alpha-globin* chains plus two identical *beta-globin* chains. Individuals with beta-thalassemia suffer anemia because their hemoglobin is short on beta chains, but—unlike another disorder affecting red blood cells, sickle cell disease, which is caused by a single mutation in the beta-globin gene—sometimes there is no mutation in a codon that can explain beta-thalassemia.

Sherman Weissman and colleagues from Yale University were intrigued by a 12-year-old Greek Cypriot girl who suffered from beta-thalassemia and was dependent on blood transfusions. They isolated and determined the sequence of the girl's beta-globin gene and solved the mystery: the mutation was not in a codon of the mRNA but rather in an intron of its gene. The mutation introduced an AG sequence that looked like a splice site, confusing the splicing machinery and causing it to go to the wrong place on the RNA. Later that year, researchers in London showed directly that the mutation in the intron caused aberrant mRNA splicing. Whenever the splicing machinery used this look-alike mutant site, it resulted in an aberrant mRNA that no longer encoded beta-globin.

Although RNA splicing can sometimes cause disease, as in these cases of beta-thalassemia, it can also be a cure. *Spinal muscular atrophy* (SMA), first described by physicians in 1890, is a devastating neurodegenerative disease that afflicts 1 in 11,000 infants and is therefore relatively common. Kids with SMA suffer from progressive weakness and loss of movement, and most pass away before reaching two years of age. The disease is genetic, caused by mutations in the *survival of motor neuron* gene number 1 (SMN1). In healthy people, the SMN1 gene makes a protein that helps snRNAs assemble with their protein partners. Because this function is essential for all cell types, it's not obvious why motor neurons should be the cell type that is particularly sensitive to loss of SMN protein, but that is the case.

Thus, the question facing any scientist who wants to cure SMA is how to compensate for the missing SMN protein. As it happens, the human genome contains many instances of duplicated genes, and there is a second survival of motor neuron gene, SMN2, which encodes the same protein as SMN1. But, significantly, SMN2 has different introns. When its mRNA is spliced, an essential chunk of codons needed to make functional SMN protein is usually skipped, leading to the production of a nonfunctional mRNA. But what if the splicing of the SMN2 RNA could be tweaked so that it could compensate for the loss of the essential protein caused by the mutation in SMN1?

In 2002, Adrian Krainer of the Cold Spring Harbor Laboratory thought of a way that the splicing of SMN2 might be rescued to compensate for the loss of SMN1. He found a spot within SMN2's mRNA that interferes with its proper splicing. Alternative mRNA splicing is often regulated by tissue-specific proteins called *splicing factors*, which bind to certain sequences on the precursor mRNA and either enhance or discourage the use of certain splice sites. These sites evolved to regulate splicing in a healthy manner, but in this case the splicing factor was interfering with proper splicing. Adrian reasoned that if this interfering sequence could somehow be covered or hidden, the correct splicing of SMN2 mRNA might be restored.

To test this idea, Adrian teamed up with a biopharmaceutical company in San Diego called Ionis, which had expertise in making drugs out of so-called antisense pieces of RNA. Such RNA is engineered to have sequences that are complementary to a given natural RNA sequence. For example, if an mRNA has a GGG codon for the amino acid glycine, the *antisense RNA* would be constructed to have the sequence CCC. Because of this complementarity, the antisense RNA will bind to the target RNA by G-C base-pairing, physically blocking a protein from binding to the target RNA.

Together, Adrian and Ionis set out to design a piece of antisense RNA that would bind to and thus "hide" the sequence that interfered with the proper splicing of SMN2 RNA and which also contained some chemical modifications that facilitated it being delivered as a drug. After several years of work, they had turned the concept into reality, rescuing the production of functional SMN protein first in cultured cells and later in mice engineered to have the same defect that causes human SMA.

But the true test would be a clinical trial conducted with real patients. Over the course of his work on the disease, Adrian had built a relationship with a Long Island family whose daughter, Emma Larson, had been diagnosed with a moderately severe form of spinal muscular atrophy, where the SMN1 gene was partially active. Emma had developed normally until age one, when she suddenly began having trouble holding her bottle or even holding her head up. Emma's parents were determined to do anything they could for their little girl who had an indomitable spirit, and they jumped at the chance to enroll her in the clinical trial of the antisense RNA drug.

Emma's mother tells the story of what happened after Emma received her second shot of the antisense drug, now dubbed nusinersen. "I was in the bedroom," Dianne Larson recalls. "Emma was in the den. Now mind you, she can't move more than a few feet. All of a sudden, I hear her voice, getting closer and closer to me. What has she done? 'Emma?' I say. Next thing I know, she's right beside me on the bedroom floor, right by the door. I was freaking out! I couldn't believe she had crawled all the way from the den."

Emma's positive response to nusinersen was shared by others in the clinical trial—so much so that the Food and Drug Administration stopped the trial a year early and approved the treatment. As of 2020, more than 8,000 SMA patients in 40 countries had been treated with nusinersen. It's not a cure—by the time the treatment takes hold, there's already some irreversible damage to neurons—but it is a lifesaver. The hope for the future is to identify newborns with the genetic defect and treat them immediately, thereby preventing SMA from taking hold.

The success of nusinersen has generated excitement more broadly about antisense therapeutics and their potential to direct RNA splicing for good. For example, Duchenne muscular dystrophy, a debilitating genetic disorder that causes a progressive loss of muscle function due to lack of a critical protein called dystrophin, might be ameliorated by changing RNA splicing patterns to generate the missing protein. Conversely, many common cancers, such as pancreatic, lung, and colorectal cancers, are driven by a pathogenic protein, such that antisense nucleic acids that inhibit mRNA splicing could potentially prevent production of the cancer-causing protein and stop the cancer in its tracks. As we'll continue to see, RNAs that have gone rogue are at the root of many human diseases. So therapies based on RNA—antisense pairing with sense—hold great promise for the future of medicine.

The discovery of splicing revealed that messenger RNA is not always a straightforward copy of the information stored in the DNA double helix. In higher organisms, including humans, the RNA that will eventually become mRNA *is* first copied verbatim from the DNA, including large interruptions in the code, the introns. But RNA splicing then chops out the introns, stitches together the coding sequences, and produces the mRNA that exits the cell nucleus to engage with ribosomes. At first glance, this process appears incredibly inefficient; what is nature up to, interrupting the coding sequences of genes with introns, only to splice them out again at the RNA level? But there's a major upside to all these gymnastics. The fact that RNA splicing can occur at alternative sites lends previously unimagined versatility to a limited genome and helps make us who we are.

The quest to understand mRNA splicing added a new bag of tricks to RNA's repertoire, as small nuclear RNAs turned out to be essential for precisely marking the splice sites. These snRNAs joined a group of so-called *noncoding RNAs*—along with transfer RNA and ribosomal RNA—that play critical roles in cellular biology. But something much bigger still awaited: as scientists would soon discover, noncoding RNAs could do much more than lay the groundwork for protein synthesis and mark sites of action. They could actually drive the action themselves. In many essential cellular processes, RNA is the catalyst.

Chapter 3

GOING IT ALONE

What turns you on? The answer, invariably, is an enzyme. These substances enable biochemical reactions in all living organisms: they make our heart beat, they break down the food in our stomach, they metabolize the alcohol we drink. Enzymes also synthesize every part of every cell in our body—from the scaffolds that hold the cell together, to the chromosomes that wrap up our DNA into tidy packages, to the greasy envelope that makes up the so-called cell membrane. Enzymes get nature's party started.

Chemically speaking, enzymes speed up, or catalyze, reactions, which sounds mundane enough—until you understand their breathtaking power. They can accelerate the natural process by which two chemicals react with each other by a factor of 10 billion. The same reaction that takes one second with an enzyme would take 317 years without it. There are roughly 10,000 enzymes in humans alone. While some enzymes are unique to our corner of the animal kingdom, many of the so-called housekeeping enzymes that keep our bodies humming are found across species, from tigers to toadstools.

Scientists as far back as the nineteenth century could see in their test tubes the result of enzymes in action. The German chemist Eduard Buchner showed that yeast cells contain an enzyme, "zymase," that converts a sugary solution into alcohol plus carbon dioxide, which comes out as bubbles. We all know this process as *fermentation*.

Very early on, scientists observed that these enzyme-catalyzed reactions were extraordinarily specific; that is, the enzymes were very choosy about what they'd react with and what products would be formed, unlike the case with chemical reactions, which are more promiscuous. Scientists measured and could predict the speed of the enzymatic reactions, for example, how the speed of fermentation would change when they added more zymase or more sugar. Entire books were written detailing the action of enzymes. Yet, quite remarkably, the scientific community couldn't decide what, exactly, these enzymes consisted of. The answer to that question became the subject of a long and contentious scientific quest. In some essential ways, that quest resembled the search for the genetic material that would eventually lead to the discovery of the DNA double helix. Just as the Augustinian monk Gregor Mendel knew that there must be some discrete unit of heredity directing the traits of his pea plants but had no idea what it was made of, scientists also knew that there was a powerful substance catalyzing biochemical reactions but couldn't agree on what it was.

James Sumner, an outdoorsman who took up chemistry after losing an arm in a hunting accident, stepped out of the scientific mainstream in the 1920s with a theory that enzymes were proteins. In his lab at Cornell University, he managed to isolate and then crystallize urease, the enzyme that breaks down urea (found in urine) into ammonia and carbon dioxide. Crystals are known to be very pure—for example, the little cubic crystals of table salt that you shake onto your burger are pure sodium chloride—so when the crystalline urease showed itself to be pure protein and retained its enzymatic activity, Sumner correctly concluded that this enzyme is in fact a protein. There were still naysayers. Yet over the next three decades, the paradigm shifted as numerous other enzymes were crystallized and also found to be proteins. By the time of his 1946 Nobel Lecture, Sumner had no qualms about stating what everyone already accepted as a fact: "All enzymes are proteins."

Three decades later, as a young scientist studying RNA, I would find myself faced with the question whether this fundamental rule might actually be fundamentally wrong. If it were, then I and other scientists would be forced to look at RNA in an entirely different way—not just as DNA's messenger, a passive player in the production of proteins, but as a catalyst that could drive biology.

IT ALL STARTED WITH POND SCUM

In 1978, after finishing my postdoctoral research at MIT, I moved to the University of Colorado in Boulder as an assistant professor. I was assigned a research laboratory on the third floor of the very dated chemistry building. My goal was to do cutting-edge science, but my lab looked like something out of the nineteenth century, with well-worn black soapstone lab benches and varnished oak drawers. Yet it was my first lab as an independent scientist, so it seemed glorious to me.

As I looked across that empty room, I had no clue what I was about to discover. But I knew I needed help, so one of the first things I did was hire a research technician. About 30 applicants answered my one-day ad in the *Denver Post*, but only one had a letter of recommendation asserting that "He has golden hands. Every experiment he does works." Art Zaug was working at Wesleyan University in Connecticut, so I interviewed him over the phone. Although I wasn't able to offer a high salary or much job security, he accepted my offer and moved to Colorado to start working with my new microscopic guinea pig—a single-celled pond critter called *Tetrahymena*.

At the time, I, along with practically every other biological scientist on Earth, still saw RNA as a kind of intermediary, always playing second fiddle to DNA. I was a DNA guy—my PhD and postdoctoral research had all been focused on the double helix. But my research was inching me closer to RNA. When I arrived in Boulder, I was trying to understand how DNA was copied to make RNA in the process called *transcription*. Just as medieval monks transcribed biblical text onto fresh parchment, cellular enzymes transcribe DNA into RNA.

The fundamentals of transcription had already been well studied in bacteria. But, as I explained to Art, I was trying to understand how transcription worked in eukaryotic organisms—living things whose cells sequester their DNA in nuclei. Most fundamental research on eukaryotes was performed using yeast, fruit flies, or mice, whose genes we can manipulate, or using human cells and tissues because of their medical relevance. I was not enamored of these options, because any given gene in a yeast or a fruit fly or a mouse is one of many thousands, a veritable needle in a haystack. I wanted to isolate an intact gene, along with its natural protein partners, so I needed a eukaryotic organism that would give me an edge.

Enter *Tetrahymena thermophila*, a single-celled furball found in freshwater ponds around the world. Shaped like a minuscule watermelon and covered in fuzzy cilia, it's cute under the microscope, a bit like a hamster without a face. *Tetrahymena* cells grow very rapidly, dividing every three hours, which means they need to double their protein content every three hours. To build the molecular factories—the ribosomes—to accomplish this feat, each *Tetrahymena* cell carries about 10,000 copies of the gene for its ribosomal RNA. Compare 10,000 copies of a gene with the typical human gene, which comes with only two copies (one from Mom, one from Dad), and you'll understand why this little guy got my attention. Easier to find a needle in a haystack if there are 10,000 needles in there. And the *Tetrahymena* ribosomal RNA genes have yet another wonderful feature: for some unfathomable reason, they exist as short pieces of DNA, instead of residing shoulder to shoulder with other genes as part of a giant chromosome. This makes it possible to isolate the ribosomal RNA genes intact, a feat nearly impossible for the thousand-times-larger human chromosomes.* It was as if the *Tetrahymena* DNA

---

*Think of the *Tetrahymena* gene as a dry piece of spaghetti, about a foot long. You could carry it down the street without breaking it quite easily. At the same scale, a typical human chromosome would be 1,000 feet long, or three football fields in length. If you had a piece of dry spaghetti that long, you couldn't even pick it up without it breaking in multiple places.

was prepackaged with a ribbon and bow, just waiting for scientists to come along and accept the gift.

JUST ANOTHER CASE OF THE BLAHS?

Our goal was to understand how these *Tetrahymena* genes were transcribed into RNA and how the proteins bound to the DNA—a special feature of eukaryotic chromosomes—might regulate this process. It didn't take long for Art's golden hands to get to work. He executed his experiments with breathtaking precision, making him a valuable resource to students in the lab. Soon they would line up to use his salt solutions—they knew that if he made them, they would work, whereas if they made them themselves, they . . . *should* work. When someone in the lab got ready to publish their results, they would sometimes ask Art to rerun a key experiment "one last time," because they knew that whereas their own experimental data would be fine, his would be flawless.

We soon found that the *Tetrahymena* gene we were studying contained a single intron, a rather small one consisting of about 400 base pairs. Our initial thought was that this was simply another case of "blah, blah, blah" interrupting the important, sensible regions of a gene, as Phil Sharp and Rich Roberts had reported two years earlier for messenger RNAs. Ours was a *ribosomal RNA* (rRNA) instead of a messenger RNA, but the basic story, we figured, would be the same. Although scientists could not agree on how these introns got into genes, we knew for sure that they had to be removed. Whenever the gene was copied into RNA, the intron had to be spliced out, very precisely, to produce the functional RNA molecule—whether it be an mRNA coding for a protein or an rRNA that would form part of the protein-synthesizing ribosome.

As a DNA guy, I wasn't all that interested in introns. I was instead striving to understand the process of transcription. Our first question would be a simple one: Could Art and I observe the RNA being copied from the DNA?

For our initial experiments, we didn't have to physically isolate the *Tet-*

rahymena ribosomal RNA genes; we instead used a little magic mushroom sauce. Art purified the *Tetrahymena* cell nuclei and added a pinch of a toxin from the beautiful red-capped *Amanita* mushroom. This ingredient certainly has no place in bœuf bourguignon, but in our biochemical recipe it had the useful property of poisoning the *RNA polymerase* enzymes that make mRNA and tRNA, while leaving the enzyme that makes rRNA unscathed. Thus, we would know that any RNA produced in our test tubes would come only from the rRNA genes. Our biochemical recipe also included a dash of radioactive nucleotides that would be incorporated into any RNA that was synthesized in the isolated nuclei, allowing us to follow it through the experiment. When everything was mixed together, we let the nuclei sit in the test tubes for an hour to give the RNA time to be made.

Art then used a technique called *gel electrophoresis* to analyze the RNA that was produced. A *gel* is a slab of wiggly Jell-O–like material, and when an electric field is applied across it—negative electrode at the top, positive electrode at the bottom—negatively charged RNA molecules are pushed through the gel. The smaller ones are able to worm their way through the gel faster than the larger ones, so the RNA molecules form discrete stripes, or "bands," indicative of their size.

Art would then take the gel into the darkroom, place a sheet of X-ray film on it, let it expose overnight, and develop it the next morning. The X-ray film worked because the RNA molecules had been tagged with a radioactive isotope, and each band of RNA slowly exposed the adjacent part of the film. So the same film used by a doctor to see if you have a broken bone could instead give us an image of the RNA in the gel.

Art and I hoped to see the ribosomal RNA, and indeed it showed up as a dark band on the X-ray film. What we were surprised to see was a much smaller RNA of about 400 bases. What could it be? A few more experiments, and Art pinned it down—it was the intron RNA, which somehow had popped out of the large rRNA transcript during our test-tube reactions.

Suddenly, our interest in the intron skyrocketed. At the time, scientists were intensely curious about *how* the seemingly extraneous introns were snipped out of RNA, and it seemed we had serendipitously captured that process in action. The first step to understanding the mechanism of any biochemical reaction is to get it to happen outside of a living organism in test tubes, where one can control all the elements of the reaction. Often it takes years to achieve such a biochemical reaction, but this *Tetrahymena* RNA splicing was happening every time we synthesized the RNA. It appeared that this RNA was giving us a front-row seat to watch the splicing performance.

We assumed that this RNA splicing was being catalyzed by protein enzymes. After all, the cutout intron had an exact length, suggesting there was a precise enzyme at work, and as Sumner had said, "All enzymes are proteins." So Art and I concocted an experiment to try to find the *Tetrahymena* splicing enzyme—or enzymes, because there could have been two: one to clip out the intron, another to stitch the useful stretches of rRNA back together. We knew *where* the RNA splicing happened—inside the *Tetrahymena* cell nucleus. And we knew *what* was being spliced—freshly copied RNA. So we first devised a way to isolate the RNA before it had undergone splicing, so that it still contained the intron. We then took test tubes and mixed two things together—the unspliced RNA and broken-up *Tetrahymena* nuclei. We ran the RNA on a gel and used X-ray film to detect any RNA splicing activity.

The first time Art tried the experiment, we were thrilled to see that the 400-base intron RNA had been spliced out of the larger rRNA. It's unusual in science to be able to re-create a natural process in an experimental setting so quickly. In fact, it took our friend John Abelson at the University of California, San Diego, four grueling years to capture the splicing of yeast mRNAs in a test tube.

But something was weird. Like any skilled scientist, Art had included a number of controls to make sure any reaction we saw was behaving sensibly. In a good control sample, one of the ingredients in a biochemical recipe is left out, leaving everything else the same. If the "experiment" were baking a cake, the controls might involve omitting only the flour, or only the eggs, or only the chocolate. The baker would soon find that the flour and eggs were essential ingredients, but the chocolate was optional, thus reinforcing his view that he knew what it took to bake a cake. In Art's RNA splicing experiment, omitting the *Tetrahymena* nuclei from the reaction was a good control, because the nuclei were the presumed source of the enzymes that would catalyze the splicing reaction. We expected this sample without nuclei to produce a whole lot of nothing. Yet, astonishingly, RNA splicing still occurred. The 400-base intron showed up bright and clear on the X-ray film, as if its splicing didn't require any enzymatic assistance.

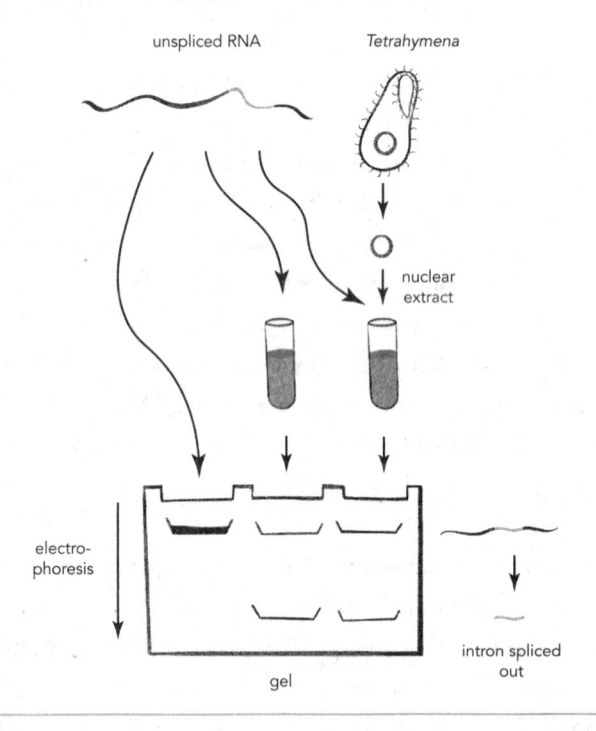

The initial experiment to look for enzymes that catalyze ribosomal RNA splicing in *Tetrahymena* gave an unexpected result. RNA splicing was expected to require the addition of an extract of cell nuclei to provide splicing enzymes, but a technique called gel electrophoresis (which separates smaller RNAs from larger ones) revealed that RNA splicing occurred both with the extract (right lane) and without the extract (middle lane). Not shown is the other product of RNA splicing, which co-migrated with the unspliced RNA.

This was not just bizarre; it was unprecedented. Open any high school or college biology textbook published at the time, and you would learn that protein enzymes were solely responsible for catalyzing reactions in cells. Yet, amazingly, here the process appeared to be happening with RNA alone. Or was it? For the next year, I remained concerned that there could be a protein enzyme from *Tetrahymena* that somehow remained *stuck* to our RNA during purification, and that this enzyme was responsible for the RNA splicing we observed in our test tubes. If our RNA was contaminated with a protein, we certainly couldn't go around proclaiming that RNA could catalyze its own splicing. Perhaps we just needed to find a way to shake the hypothetical protein off, and then our RNA would quit splicing itself and we could return to looking for the splicing enzyme.

At that juncture, I did what any good scientist would do: I followed the data. Its trail soon led me into what was then the small world of RNA science. The DNA guy would need to become an RNA guy. Though I didn't realize it at the time, jumping ship would turn out to be the most momentous decision of my life.

LITTLE DONUTS OF RNA

While Art and I continued to explore the splicing reaction, the new grad student in our lab, Paula Grabowski, made yet another unprecedented finding. We had been doubting that RNA could really splice itself, and now Paula added another equally strange result to the mix. It may seem counterintuitive, but somehow having *two* observations that seemed to come from outer space gave us comfort that we were not incompetent—or crazy.

Paula hadn't set out to make a new discovery; rather, it found her. She had decided to run the RNA splicing reaction at 39°C instead of our standard 30°C, both temperatures being within the growth range of *Tetrahymena*. Now instead of a single cutout intron product, she saw two. And there was something weird about the new intron species. When analyzed by gel electrophoresis, the new intron moved very slowly. The RNA seemed to have an unusual shape that was slowing its movement through the gel.

Although circular RNAs were rare and thought to be restricted to viruses and virus-like infectious RNAs, the behavior of this intron during electrophoresis certainly smacked of a circle or perhaps a branched structure. Paula had developed several lines of evidence suggesting that the new intron species was indeed a circle. But we would not know for sure until we could catch a glimpse of the RNA through an electron microscope. Because I was the only electron microscopist in our lab, this task fell to me.

It had been two years since I'd set up my lab and hired Art. I was teaching full-time then, so I had to do my own lab work largely at night. I had mounted Paula's RNA samples on grids, disks about an eighth of an inch across containing a crisscross latticework of copper, and I booked time that evening on one of the campus electron microscopes. As I turned up the high-voltage dial in the dark room, the fluorescent screen lit up my face with a Halloween-green glow. Peering through the binoculars, I was thrilled to see little donuts—tiny circles of RNA—covering the grid. Seeing is believing! Admittedly, a circular intron RNA seemed more of a curiosity than a breakthrough at the time. It would take more work before we could understand its significance.

That summer, I was invited to give a talk about our new RNA work at a prestigious conference on nucleic acids being held in New Hampshire, a rare opportunity for an assistant professor. I always telephoned the lab to try to help keep the research moving while I was traveling. When I spoke to Paula the day after I arrived, I noticed the undercurrent of excitement in her voice. Paula had isolated the linear form of the *Tetrahymena* intron, after it had been spliced out of the larger RNA, and she found that as it sat around in the test tube, it began to convert itself to the circular form. In the history of RNA research, getting an RNA to form a circle always required adding a protein enzyme to join the two ends of the nucleic acid—circularization simply didn't happen spontaneously.

As much as I tried to share her excitement, I was mostly incredulous—and even a bit annoyed. Why was an inexperienced graduate student bothering me with something that was totally impossible, clearly some outrageous mistake, just before I was to speak to this esteemed group of scientists? No way was I going to mention these bizarre new results in my talk the next day.

Yet after I got back to Boulder, I saw that Paula's impossible experimental result was in fact true. It was totally reproducible. So now my young lab had found two wacky results that violated everything in the textbooks: RNA seemed to be splicing itself, in the absence of any likely source of an enzyme, and the cutout intron was tying itself into a circle, again in the apparent absence of any enzyme. What was happening?

By the end of 1981, we had dissected the RNA splicing and circularization reactions down to the level of individual phosphorus and oxygen atoms. But regarding the source of catalysis, we were in exactly the same situation as enzymologists were in 1917, before James Sumner. We knew what was happening, but we remained befuddled about what was causing it to happen. RNA would not cut and then join itself together spontaneously, unless you waited a few thousand years. Even then, the reaction wouldn't occur with the pinpoint accuracy we were seeing. There had to be a catalyst.

As it would turn out, it was staring us right in the face.

PLUCKING THE DAISY PETALS

At the chemistry department Christmas party that year, Paula presented me with a little handmade gift. It was a plastic daisy flower, on which she had meticulously printed, on alternate petals, "it's a protein" and "it's not." That was our conundrum.

The RNA we were using for our experiments had been put through

Picking the petals off the daisy flower was not the way to resolve the question whether *Tetrahymena* ribosomal RNA splicing required a protein enzyme.

rigorous purification steps to remove protein. The well-known techniques for removing protein from RNA are not so different from what we do to remove tough stains from our clothes. We wash them in hot water, because high temperature unravels and inactivates protein chains. We use detergents, which also unravel protein chains. We might even treat our clothes with an "enzyme-active laundry detergent," because there are well-known protein enzymes that chew up any other protein they encounter. Our RNA had been put through all these treatments—and it kept on splicing and then converting itself into a circle. The evidence simply wasn't supporting the hypothesis that our preparations contained a contaminating protein. Yet I knew that if we claimed there *wasn't* a protein—that the RNA by itself was capable of all these acrobatics— then some skeptical scientist was sure to remark that he or she had once heard of a protein that could survive all of the insults we used in our purification scheme.

What we needed was to produce the intron-containing unspliced

RNA without using *Tetrahymena* so that we could safely discount contamination by *Tetrahymena* enzymes. If *artificially* produced RNA still underwent splicing, at the same sites known to be used for splicing in the living cells, then we could proclaim that the RNA was its own catalyst. If that was true, it would revolutionize our understanding of not only what enzymes can be made of but also what RNA is capable of.

These were still early days for genetic engineering, and my lab was among the many that had not yet mastered this technology. The process we needed to use was the same one that biotech companies use today to engineer genes when they create new drugs. We had to trick *E. coli* into making the *Tetrahymena* ribosomal RNA gene. This required inserting the *Tetrahymena* DNA into a *plasmid*, a circular DNA that would be replicated in the bacterial cell. By doing this, we essentially turned the *E. coli* in our Petri dishes into genetic photocopiers, spitting out as much of the desired gene as we needed for our work.

This type of experiment, which would now take an undergraduate in my lab a day to do, took us many months of fumbling in 1982 before we perfected the process. Once we finally had our squeaky-clean gene that had never been exposed to a living *Tetrahymena* cell, we still needed the protein enzyme called RNA polymerase to copy that genetic material into RNA. I was fortunate that my wife, Carol, was an expert at purifying *E. coli* RNA polymerase, and she provided this final ingredient. (It sometimes helps to be married to a fellow biochemist, although I'm sure she would've given me a drop of this stuff even if she wasn't my wife.)

After Art Zaug copied the artificial gene into RNA, he next used well-honed procedures to remove the one protein we had put into the experiment—the *E. coli* RNA polymerase. He then ran the purified RNA through the splicing reaction, putting all the necessary components into little plastic tubes at different concentrations. Once again, it's a bit like cooking. To bake a cake, you have a recipe that calls for flour, sugar, eggs, baking powder, and water. In our case, the recipe called for the RNA, some salts that are common to all cells, and guanosine—one of the four building blocks of RNA, the G nucleotide in the RNA alphabet.

As before, the products of the RNA reaction were separated by gel electrophoresis and visualized on X-ray film.

My friend Jan Engberg, a professor from Copenhagen who had helped introduce me to *Tetrahymena* a few years earlier, happened to be visiting us in Boulder when we carried out this experiment, testing whether the artificial bacterial version of the *Tetrahymena* RNA could still perform its magic. I still remember Jan's reaction when he saw the newly developed film. The intron had spliced itself out of the larger RNA, forming a sharp 400-base product on the gel. This time we knew for certain there was no protein enzyme involved. His voice growing thick, Jan looked up and said, in his wonderful Danish accent, "You've done it."

Now it was time for some fun. What should we call this remarkable RNA? I set aside an area of the lab chalkboard for the naming contest, and as the week went by, more and more entries appeared. Among them was the inevitable "sex RNA," standing for "self-excising RNA." Also "Circulon," a new superhero capable of converting himself or herself into a circle. Then there was "ARNzyme," using the French abbreviation for RNA, ARN (Acide ribonucléique), to create a portmanteau that would roll off the tongue more easily than RNAzyme. But one entry stood out: *ribozyme*, a ribonucleic acid with enzymatic activity. The fact that the RNA not only spliced itself but also welded itself into a circular form convinced us that it was acting like an enzyme, as it had enough power to keep chugging along even after accomplishing the splicing reaction.

Certainly, it was bold to adopt such sweeping terminology. We had but one example, and we had chosen a name that would fit an entire class of molecules. But I saw little risk. If no second example were ever found, our discovery would remain a quaint outlier, and it wouldn't matter much what we named it. But what if *Tetrahymena* had provided the world with the first example of a large class of active RNA molecules?

It didn't take long for that question to be answered in the affirmative. Within months of our December 1982 publication announcing self-splicing RNA, I heard from colleagues in St. Louis, Amsterdam, and Albany who had discovered intron ribozymes in a fungus, baker's yeast, and even a bacterial virus—the last of these breaking the "rule" that RNA splicing was a strictly eukaryotic phenomenon. Once scientists knew that self-powered RNAs could exist, they started finding them throughout nature.

Self-splicing RNAs seemed to smash what had been one of the basic tenets of biology—that all enzymes are proteins. It wasn't that Sumner was so far off base. *Most* enzymes are indeed proteins. Nevertheless, all these self-splicing RNAs sparked speculation that there may have been an ancient time, before the advent of proteins, when ribozymes ruled catalysis.

The discoveries also sparked speculation whether there might be yet-undiscovered RNA catalysts out there in nature, performing all kinds of dazzling reactions that were previously thought to be the sole preserve of proteins. And indeed, it turned out that a very different sort of RNA catalyst was just around the corner, waiting to be discovered.

A GRAIN OF SALT

Unlike the goddess Athena, who sprang fully formed and even fully armed from Zeus's head, RNA molecules are not transcribed from DNA in their final active form. Instead, they undergo processing before they are put to work. Splicing—in which introns are lopped out and the remaining RNA sequences are sewn together—is only one, admittedly dramatic, type of *RNA processing*. Other RNA-processing events take place at or near the ends of freshly made RNA, and they include cutting off unneeded sequences as well as adding bases not encoded by the DNA.

Transfer RNA, the adaptor that recognizes an mRNA codon at one end and carries the matching amino acid at its other end, is a good example. It is initially transcribed with extra RNA at the beginning of the adaptor, and this appendage must be cleaved off—precisely at a certain base—for the tRNA to be functional. An enzyme that cuts RNA is called a *ribonuclease*, abbreviated RNAase or RNase, and the particular enzyme that cuts the unwanted extra sequences preceding a tRNA is called *ribonuclease P* (RNase P). P is for processing. RNase P was dis-

RNase P

tRNA precursor

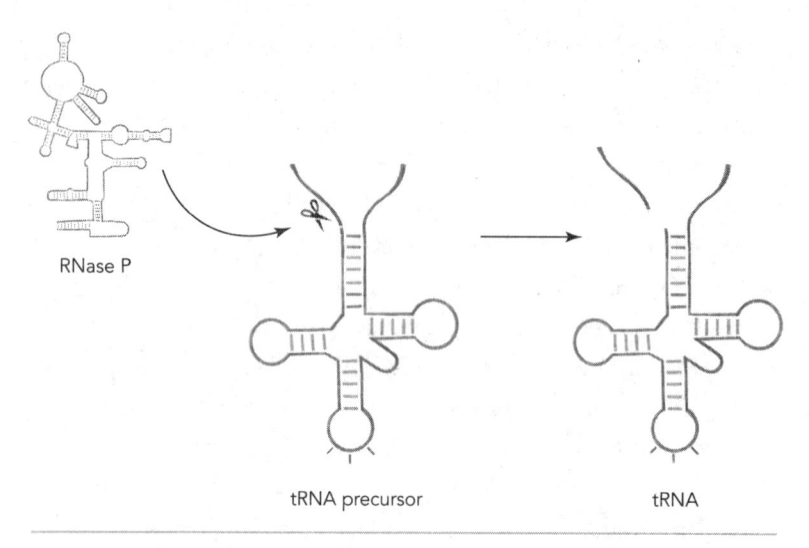

tRNA

Ribonuclease P, which in *E. coli* consists of an RNA molecule (shown here) and an associated protein (not shown), cleaves a specific site of the tRNA precursor molecule to create the correct end of the tRNA.

covered in *E. coli* by Sidney Altman when he was a postdoctoral fellow in Cambridge, England, in the same laboratory that gave us Francis Crick and Sydney Brenner.

Later, in his own laboratory at Yale University, Sid Altman continued to investigate RNase P. And a curious enzyme it was. Whenever it was purified from *E. coli* by using the techniques that had been developed over many decades for protein purification, a pesky RNA molecule came along for the ride. Ben Stark, Sid's graduate student, was under pressure to purify RNase P for his PhD dissertation, and he endured a lot of ridicule for not being able to remove the RNA from his preparations of the enzyme. After all, ALL ENZYMES ARE PROTEINS, and a competent graduate student should be able to purify the RNase P enzyme to an RNA-free state. But Stark was strong enough and skilled enough to reject the idea that he was incompetent, and he eventually did experiments that convinced Sid and his dissertation committee that indeed the RNA component was essential for RNase P enzymatic activity. When they purified the RNA and protein components separately, they needed to mix them together again to restore tRNA-cleaving activity. Yet, they didn't think they were rewriting so much as tweaking the golden rule about enzymes: they still thought that it was the protein component of their system, and not the RNA, that must be making the reaction happen.

So why did RNase P need protein *and* RNA? It took a dose of serendipity to sort things out. The Altman lab was collaborating with Norm Pace's lab in Colorado on a series of mix-and-match experiments. Sid had the purified RNase P protein and RNA from *E. coli*, and Norm had the two components from a distantly related bacterium, *Bacillus subtilis*. They were curious as to whether either of the cross-species combinations, such as *E. coli* RNA plus *B. subtilis* protein, would be enzymatically active. The lightning-bolt experiment was performed on Friday, September 23, 1983, by a staff scientist in Sid's lab, Cecilia Guerrier-Takada. As an excellent scientist, she included several control experiments alongside the mix-and-match combinations, just like Art had done in my lab. Critically, she retested the RNA-only and protein-only reactions, which had previously been negative. She was confident that they would again give no activity.

But something was different this time around. Norm had suggested to Sid by phone that it might be useful to add extra magnesium chloride, a common salt found in all living cells, in a set of reactions. It's what many cooks do when they see a recipe in a cookbook: they alter it a bit, testing whether they like the cake better if they add a couple more eggs or a bit less sugar. So Cecilia incorporated these new conditions along with her standard conditions when she mixed her tRNA precursor with the various RNase P RNA and protein components. She then used gel electrophoresis to separate the tRNA precursor from any reaction products, put on an X-ray film, and let it expose overnight.

When she developed the film on Saturday, she saw that the tRNA precursor had been trimmed at the correct site by both RNase P enzymes, the *E. coli* and the *B. subtilis* versions, as always happened. She'd seen similar results dozens of times. Then, *Voila!* The test-tube solution containing only the RNase P RNA—without its protein partner—and the big dash of magnesium salt yielded the precise trimming of the tRNA precursor. The RNA by itself appeared to be acting as an enzyme, while the protein-only samples remained inactive.

Immediately understanding the groundbreaking implications of her result, Cecilia wanted to make sure it was correct before telling anyone. She set up a repeat of the experiment that same Saturday, to make sure that she hadn't, for example, mixed up the test tubes. She developed the X-ray film on Sunday, and indeed, both the *E. coli* and *B. subtilis* RNA subunits were again active as enzymes under the high-salt conditions, whereas the protein components showed no activity without the RNA. So much for "All enzymes are proteins!" Sid was in his office on Sunday, so Cecilia was able to show him the astounding results and share his delight at the discovery.

When he came into the lab on Monday morning, Sid had already written a rough draft of a scientific publication announcing the discovery. They telephoned Norm's lab to share the findings with their coauthors. Norm was equally surprised; the axiom that "All enzymes are proteins" had been cemented in his worldview, as well.

The discovery that RNase P was a ribozyme, coming a year after we reported our self-splicing RNA, extended the concept of ribozymes in an important dimension. In our experiments, we had discovered an RNA that could splice itself—that could be its own *internal* catalyst. Now Sid's team had found that RNA could work as an *external* catalyst, acting upon something else—tRNA precursors. In both cases, RNA had emerged as a molecule that was not merely carrying information from DNA to protein but was an active driver of cellular reactions. And that is why, six short years later, Sid's and my complementary discoveries were recognized by the 1989 Nobel Prize in Chemistry.

BREWING UP MORE CATALYTIC RNAS

Science works in mysterious ways. You make a hypothesis, gather evidence, run experiments, check your data. If you're fortunate, you make a discovery that your peers regard as a meaningful contribution to the field. But you cannot predict what will happen next, who will take up the baton, and where they will run with it. In the case of ribozymes, RNA catalysts next popped up in Australia, hiding in plant infectious agents with names such as avocado sunblotch viroid. These "hammerhead" ribozymes catalyzed a very simple reaction—not pounding a nail but rather cleaving themselves or other RNA molecules at a specific site. They attracted much attention because of their small size, about 30 nucleotides. So it didn't take an enormous RNA to act like an enzyme.

Scientists soon found that the catalytic power of RNA had also been hiding in the snRNAs involved in mRNA splicing. It turns out that nature relies on snRNAs for much more than simply marking where the splicing should happen, as important as that is. Beyond that, a group of snRNAs collaborate to catalyze the cutting and joining reactions required for mRNA splicing. Many scientists studying many biological systems contributed to unraveling these roles of the snRNAs, among them Christine Guthrie of the University of California, San Francisco.

The particular system Christine used may seem unlikely, as indeed it did to many other RNA scientists at first: the yeast that we use to brew beer. But yeast has astoundingly facile genetics—that is, it is easy to mutate a yeast gene and see what happens as a result—and it therefore promised to provide insights about the fundamental mechanics of splicing, or so thought Christine in 1980. She would often extol "The awesome power of yeast genetics." If she could find the yeast snRNAs that corresponded to the mammalian U1, U2, U4, U5, and U6 snRNAs—the ones that by then were thought to be essential for mRNA splicing—she believed that she could engineer small changes in their sequences and figure out whether their bases were pairing with the mRNA or with other snRNAs. If she pulled it off, she would help unravel the mechanism behind the cutting and joining reactions that make mRNA splicing work.

It was fortunate that Christine had such a passion for research, because she often received faint encouragement. Her graduate school advisor told her, "Girls can't do biochemistry. They can't lift heavy rotors [that hold test tubes in a centrifuge] or spend long hours in the cold room [the walk-in fridge used to perform biochemical purifications]." When she struggled for years to identify the yeast snRNAs—after all, you can't mutate a gene until you've identified it—many in the RNA research community were disparaging. They repeatedly pointed out that because very few yeast genes have introns, the yeast mRNA splicing mechanism might be unique and unrelated to that of humans. In retrospect, it seems very odd that some biologists, who live and breathe evolution, would doubt that fundamental aspects of splicing would be conserved among those species that had introns. But the difficulty of the hunt for yeast snRNAs fueled their skepticism. Was it possible that yeast didn't have snRNAs?

Christine and her lab spent five long years combing through yeast cells before they finally found their first snRNA, the one called U5. Using the bag of tricks known to yeast geneticists, they showed that cells depleted of this RNA stopped growing and accumulated unspliced RNAs—indicating that yeast U5 must be involved in the splicing process. Other research groups joined the hunt, and soon more snRNAs were found that were responsible for different steps of the splicing reaction.

Christine now had identified the different pieces of the splicing mechanism, but she saw that they could fit together in more than one way and that their position changed over the course of the reaction. This was not the kind of problem one could solve overnight.

By 1986, some progress had been made. Researchers confirmed that U1 bound to the left end of each intron, as Lerner and Steitz had originally proposed, but later it left the scene. U2 bound near the right end of each intron, as demonstrated by a student in Christine's lab, Roy Parker. What happened after that, however, remained a mystery.

Around that time, I was one of a group of RNA scientists who gathered in the Rocky Mountains for an annual skiing-and-cooking adventure that Tom Steitz dubbed "RiboSki." After a day in the snow, Christine, Joan Steitz, and John Abelson would spend hours comparing RNA splicing data, while Elsebet Lund and Jim Dahlberg of the University of Wisconsin, Olke Uhlenbeck from Boulder, and I looked on. We all wondered how the snRNAs might snuggle up to each other at different stages of the mRNA splicing reaction. I'm sure that the convivial atmosphere produced many good ideas, but it would take Christine several more years of dogged research before she was able to map each step of the splicing reaction.

A breakthrough came one night in 1992. She was working late in her San Francisco lab, pondering new results she'd obtained with her student Hiten Madhani, concerning the snRNA called U6. That turned out to be the key. As she sketched the solution to the splicing mechanism, it looked less like a biochemical reaction and more like an intricately choreographed ballet. U6 enters the scene tightly entwined with U4, and together they find their place on the intron. But then U2 butts in, stealing away U6, and U4 exits the stage in a huff. Now U6 and U2, with some help from U5, are free to produce the chemistry needed to complete the splicing reaction. Together they act as a ribozyme, catalyzing the mRNA splicing process.

Christine was so dazzled by what she had found that she just had to tell someone, but the building was almost dark. Emerging into the hall, she found a janitor sweeping the hallway and shared her new insights with him.

"And he kind of got it!"

Why did so many scientists—initially even those in my and Sid Altman's and Norm Pace's labs—pooh-pooh the idea that RNA could be an enzyme? Why were we tied so tightly to the idea that a protein must be at the heart of catalysis? In part, it was because we knew that protein enzymes fold into intricate shapes that are custom-made for the tasks they need to perform. Disrupt that structure by boiling the proteins or by genetic mutation, and the activity is lost.

By contrast, we didn't know enough about RNA structure at the time to see how RNA could fold itself into a catalyst. Messenger RNAs in particular were always pictured like strands of cooked spaghetti, highly flexible and not forming any stable shape. Even if the spaghetti was looped and twisted on your plate, it straightened out as soon as you picked it up with a fork.

With RNA, scientists were thinking in one or two dimensions. We were thinking of a linear order of A, G, C, and U, lined up like the letters in this sentence, and about the pairing of bases within the sequence, as in the U1 snRNA structure illustrated above. But to appreciate how it's possible for RNA to be an enzyme, one needed to see it in three dimensions. Remember how little we knew about the molecular mechanisms of genetics until we saw the three-dimensional shape of DNA, the double helix. Similarly, we could not hope to understand catalytic RNAs until we solved their various structures.

Chapter 4

THE SHAPE OF A SHAPESHIFTER

"Form follows function" is a famous axiom in architecture, but it holds for pretty much all of the physical world. A hammer and a screwdriver have different shapes, each attuned to its use, but both have a similar handle because that part of their function—fitting into a human hand—is the same. The same principle applies at the cellular level. If a protein enzyme works to break down food into small pieces that can be metabolized, then the enzyme has a cleft in it to accommodate food molecules, such as potato starch, that need to be broken down. If a protein's function is to move a muscle, then it needs a stretchy region that can expand and contract.

This relationship between form and function means that we can't really understand the molecules of life until we know their structure, how they are built and how they fit together. Without that structure, researchers in the life sciences are a bit like mechanics trying to repair an automobile engine in total darkness—the process would be very slow, inefficient, and frustrating. Having a structure is like turning on the lights; now the mechanics can see all the parts of the engine, how they fit into each other, what part or connection is defective, and how to fix it.

The first generation of molecular biologists saw the deciphering of the physical structure of proteins and then of DNA as grand and worthy challenges. The technique they used is called *X-ray crystallography*. The method works by shooting a beam of X-rays at a sample, such as a crystal of a protein molecule, collecting images of the diffracted radiation, and back-calculating what the structure must have been to produce that diffraction. Consider a pebble thrown into a still pond. It produces an emanating series of waves that could be used to determine the exact place where the pebble hit the water. Now consider throwing a whole handful of pebbles into the pond. The wave patterns are much more complex and overlapping, but the information about where each pebble landed is still there. Similarly, the diffraction pattern produced by an X-ray beam aimed at a protein crystal can reveal the location of individual atoms in the protein.

How about RNA? We have seen that RNA has wondrous functions, and we will encounter many more. Each of these functions must have a corresponding form, a specific structure that enables it. But the shape of RNA proved to be a lot trickier to identify than DNA's.

Jim Watson found this out the hard way. After codiscovering the double helix, Watson thought he would solve the structure of RNA as an encore. But he ran into a problem. DNA has only one shape, the double helix, in which each strand pairs with its sister strand. This twisted ladder keeps both strands locked up and in check—or, as we RNA scientists sometimes joke, it keeps the two DNA strands from doing anything very interesting, such as catalysis. RNA, conversely, doesn't have one shape but *millions* of possible shapes. Freed from the constraints of a double helix, RNA can assume a virtually limitless number of forms, which accounts for its stunning versatility. The fact that RNA is a changeling, however, makes it all the more important to understand the various poses it strikes. Yet mapping this shifty substance has been notoriously difficult.

Watson struggled with RNA structure for a decade. He initially purified RNA from diverse sources including plant viruses, calf liver, and yeast; performed X-ray diffraction experiments; and—on the basis of very crude data—concluded that these diverse RNAs had a single, common structure. This was rather like looking at an elephant and a Volkswagen from a distance of 200 meters on a very foggy day and concluding that they are the same. If you had a pair of binoculars and waited until the sun came out, you would come to a very different conclusion.

Before relenting, Watson actually took a step in the right direction. He moved from studying a diverse collection of different RNAs—which had multiple functions and therefore multiple structures—to purified ribosomes, the RNAs of which have specific structures attuned to the specific function they perform in protein synthesis. But the technology and know-how to solve ribosome structures would take another 40 years to develop.

Yet once we finally had complex RNA structures in front of us, we could directly see how RNA works its magic—how it serves as a machine to construct essential protein molecules, or builds out the ends of our chromosomes, or precisely edits the DNA in human cells. All these major discoveries about the catalytic power of RNA lay in the future. At the beginning, success came by starting small.

BABY STEPS

Bob Holley of Cornell University picked up where Jim Watson left off. In the late 1950s, he realized that it was folly to try to divine a single structure from a mixed batch of RNAs, so he zeroed in on tRNA, the adaptor that connects an amino acid with its triplet codon. Transfer RNA was small enough to give him a chance of determining its nucleotide sequence—something that at the time had never been done for any type of RNA.

Why did Holley have to sequence the RNA before he or anyone else could take a stab at determining its structure? Solving an RNA structure is a bit like diagramming a sentence. Even if you were the world's greatest grammarian, you couldn't hope to diagram a sentence if you couldn't read that sentence first. The nucleotide sequence gives you the order of the chemical bases A, U, C, and G—the letters in the words of your sentence—and once you see them all strung together, you can begin to diagram your molecule, figuring out how these elements are positioned in space, how they work together.

For a source of tRNA, Holley chose yeast—the same yeast we use to bake bread and brew beer. He knew that tRNAs were relatively abundant in yeast, and he could buy as much Fleischmann's yeast as he needed from a local bakery. Yet it would take three years and 300 pounds of yeast to purify 1 gram (about the mass of a raisin) of one type of tRNA.

The tRNA that Holley was able to separate from the rest happened to be the one that served as the adaptor for the amino acid alanine. Once he and his research team had isolated it, they began to chop it into pieces that were small enough to analyze chemically and then to figure out how those pieces were ordered. Within a year they had deciphered the nucleotide sequence, paving the way for them to solve the first structure of an RNA.

In 1965, Elizabeth Keller, a seasoned research scientist in Bob Holley's team, took up the challenge of predicting how the alanine tRNA might fold onto itself in two dimensions. She knew the sequence of bases, but how might they interact?

Remember that RNA is usually single stranded. In DNA, the bases from one strand pair with those of the other to form the rungs of the DNA ladder, creating a similar double-helical shape no matter what the sequence. But with RNA, the sequence determines the shape, as bases from one part of the strand find bases in another part with which to pair. A single G-C pair is too weak to hold together, but—to give one example—if there are four consecutive G's that can pair with four consecutive C's, then the four base pairs hang together. These pairings, determined by the sequence of the RNA, cause the RNA to fold back on itself, creating "hairpins," branches, loops, knots, and a myriad of other possible shapes. Keller quickly saw that she could form base pairs in many different combinations to fold up the tRNA. Which was the right one?

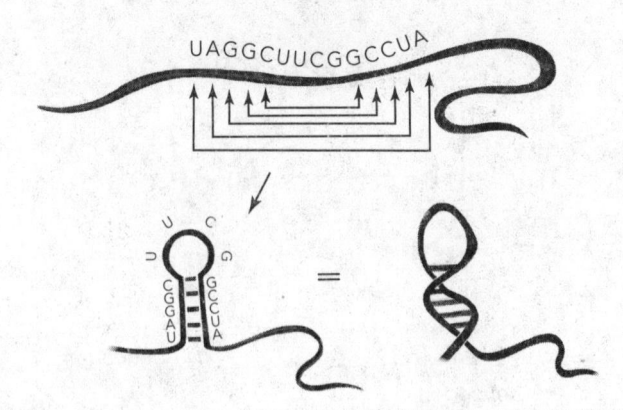

Single-stranded RNA can form base pairs within the molecule to give shapes such as the "hairpin," or stem-loop structure, shown here. At the bottom, one representation makes it easy to see which bases are paired (left), while the other shows the 3D shape of the hairpin (right).

One clue involved the three bases of the tRNA that provided the connection with the mRNA codon. With a sequence complementary to that of the codon, this triplet is called the anticodon. It seemed reasonable to Keller and Holley that this anticodon would not be buried within the folds of the tRNA structure but would stick out in order to pair easily with the mRNA.

Keller gathered pipe cleaners and pieces of Velcro to model different base-pairing possibilities. She settled on a distinctive three-leaf clover pattern, which met the expectation of having the anticodon unpaired in a loop that capped off the middle arm of the structure, ready to pair with the corresponding mRNA codon.

Soon sequences of a dozen other types of tRNA were determined, and each and every one could theoretically fold into the cloverleaf structure. Because all tRNAs needed to fit into the same slots in the ribosome to deliver their amino acids for protein synthesis, they would need to have the same shape. So seeing that "one shape fit all" provided great support that the cloverleaf was correct for tRNA.

Though it was clearly a major breakthrough, Keller's cloverleaf model had one big shortcoming: it lay flat on the tabletop, showing tRNA in only

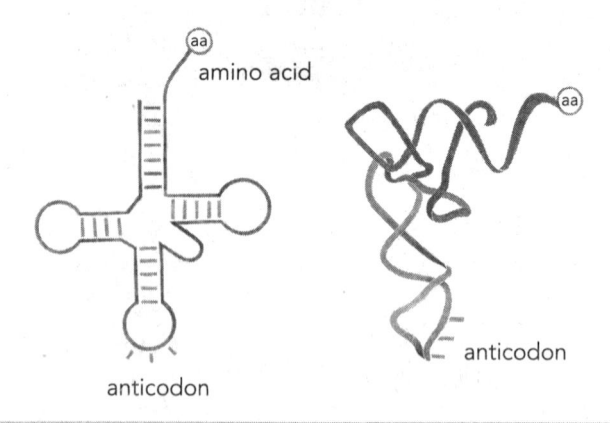

Pairing of bases gives a cloverleaf shape to a tRNA molecule (left), which then further folds to give an L-shape in three dimensions (right). The three-base anticodon is exposed, allowing it to pair with an mRNA codon and bring the correct amino acid (aa) into the ribosome.

two dimensions. I sometimes refer to such two-dimensional representations as "roadkill," because they show us what RNA would look like if it were run over by a truck. Just as it would be difficult to understand squirrel behavior from analyzing a flattened squirrel, one can't really understand RNA behavior from a 2D model.

In the late 1960s, a race broke out to determine the three-dimensional structure of tRNA. It was Cambridge versus Cambridge: one team was led by Sung-Hou Kim and Alex Rich at MIT and the other by J. D. Robertus and Aaron Klug at the Laboratory of Molecular Biology in the other Cambridge.

The method of choice was X-ray crystallography, which was by this time yielding protein structures left and right. Growing crystals in the lab involves making a very concentrated solution of the purified molecules and then setting up rows of drops that contain different concentrations of additives such as salts. You're looking for the Goldilocks conditions that give suitable crystals. If the molecules are too soluble in a given drop, then the solution stays clear—no crystals. If the molecules are too insoluble in another drop, then they precipitate out of solution into a useless lump. But in the occasional drop, where the solution is right on the edge between soluble and insoluble, the molecules snuggle up next to each other in straight rows and columns. Peering through a microscope, you see beautiful sharp-edged crystals growing slowly and getting larger day by day. Then you mount a single crystal in front of an X-ray beam, turn on the radiation, and collect the pattern of diffracted X-rays. After a few more tricks and a lot of computation, the structure can be "solved," by which we mean that we get a model of where every atom of the molecule is positioned in 3D space.

Crystallization is half science, half art, so the researchers learn to take what they can get. The tRNA that crystallized most easily was phenylalanine tRNA, so that was the one that both competing teams solved. Its structure showed that the cloverleaf envisioned by Elizabeth Keller was a reality, and the cloverleaf was further folded on itself to give an L-shaped molecule. At one end of the "L" was the triplet anticodon, and at the other end was the corresponding phenylalanine amino acid.

Determining the first 3D structure of an RNA with a known biological function was a thrilling achievement—one shared by the two Cambridge teams in 1974. In such cases, the first breakthrough is often followed in quick succession by many more. The first domino falls and starts a cascade. But this was not the case for RNA structure. In the 15 years after phenylalanine tRNA, no RNA larger than tRNA had its structure solved. Despite the extensive efforts of countless researchers, all the other known types of RNA proved too slippery to pin down. And without a structure to act as a guide, figuring out the roles of individual nucleotides in a large RNA—such as the *Tetrahymena* ribozyme—is a painstakingly slow and tedious process.

THE BUTTERFLY CATCHER

While some were waiting around for the technology to get good enough to crystallize more complex types of RNA, François Michel started dreaming. François was a researcher at the French National Center for

Scientific Research (CNRS) in the town of Gif-sur-Yvette, outside of Paris. One of his passions was collecting, breeding, and understanding the genetic basis for different species of butterflies, and he collected RNA sequences just as avidly. He had an astounding memory for these sequences. It's said that he would compare them and try fitting them together in various arrangements in his head, even while he slept. He looked the part of an eccentric genius—huge head of hair, big beard. When I saw him at conferences, I sometimes thought he had just emerged from the forest after a months-long butterfly hunt.

François's colleagues in Gif had identified a new group of introns in yeast mitochondria, the energy-generating part of the cell, that had fascinating genetic properties. They noticed that little scraps of almost identical sequence were sprinkled in nine different introns. From one yeast intron to another, the scraps resided in the same order, suggesting they had a similar function. François knew that these similar bits of sequence were vital to the RNA splicing reaction because, when mutations arose in these scraps of RNA, the splicing reaction wouldn't work. Furthermore, pairs of these sequences were complementary to each other, suggesting that they zipped up to form stem-loop structures (as in the illustration above). By 1982, François proposed how these paired-up RNA sequences could form a similar 2D shape for all of these yeast mitochondrial introns.

But why would introns need to form a specific shape? After all, if they were similar to the mRNA introns that Phil Sharp and Rich Roberts had discovered, the intron RNA might need to be unstructured so that it could pair with the U1 and U2 snRNAs. The mystery of the structured introns didn't last long, because shortly after François proposed his structural model, we reported our discovery of the ribozyme in *Tetrahymena*. It took François only a second to look at the nucleotide sequence of the *Tetrahymena* intron and realize that his 2D model would also work for our self-splicing intron. This was remarkable and unexpected—yeast are evolutionarily very distant from ciliated protozoa such as *Tetrahymena*, and mitochondrial genes are very different from genes in the cell nucleus. So why would these otherwise unrelated RNAs fold up into the same shape? The answer, totally unprecedented, must be that this shape was necessary for a self-splicing catalyst, and the yeast mitochondrial introns must also be self-splicing. Indeed, by 1985 a Dutch group confirmed this prediction.

François's 2D models looked somewhat like expanded versions of the tRNA cloverleaf. The intron RNAs were several times larger than a tRNA, and their structures were also more elaborate, featuring a dozen or more stems and hairpin loops instead of the four found in tRNAs. Having this road map to the structure of a class of ribozymes was a substantial achievement, but François knew that RNA catalysis did not take place in two dimensions. He longed to build a complete 3D model based on sequences alone, something that no one had ever done for a large RNA.

In 1983, François met Eric Westhof of the University of Strasbourg at a scientific conference. Eric had spent his boyhood in the Belgian Congo, got a degree in physics from the University of Liège in Belgium, and later studied tRNA structure using X-ray crystallography at the University of Wisconsin. Eric's training and skills complemented François's perfectly, and they shared a passion for turning 2D models of RNA into 3D reality.

Strasbourg lies in the vineyard-rich Rhine valley, not very near to Paris, so François would bring his sleeping bag when he visited Eric for model-building sessions. Eric would sit at his computer, constructing RNA helices that corresponded to known base-paired regions of the *Tetrahymena* intron structure. That part wasn't difficult—RNA double helices are like little snippets of the DNA double helix, and the details of their helical angles would follow those seen in the tRNA crystal structures. The hard part was figuring out how these little helical units came together in 3D space to construct a catalytic shape. Hopefully the sequences of the RNA connecting the helices would provide clues about their 3D arrangement, just as such connecting sequences serve to fold the tRNA cloverleaf into its L-shaped 3D structure.

François would sit next to Eric, interrogating printouts of 87 sequences of related introns, all of the self-splicing type. These printouts supplemented the many sequences he carried around in his head. As Eric moved parts of the RNA into new arrangements on his computer, François would look for any coordinated changes in the sequences of different introns that hinted certain nucleotides were touching each other in 3D space. When François saw such evidence, he'd proclaim, "That works!" At night he'd curl up in his sleeping bag next to Eric's computer, visions of RNA sequences dancing like butterflies in his head.

By 1990, François and Eric had their 3D model of the *Tetrahymena* intron—and to an RNA biologist, it was a thing of beauty. It looked rather

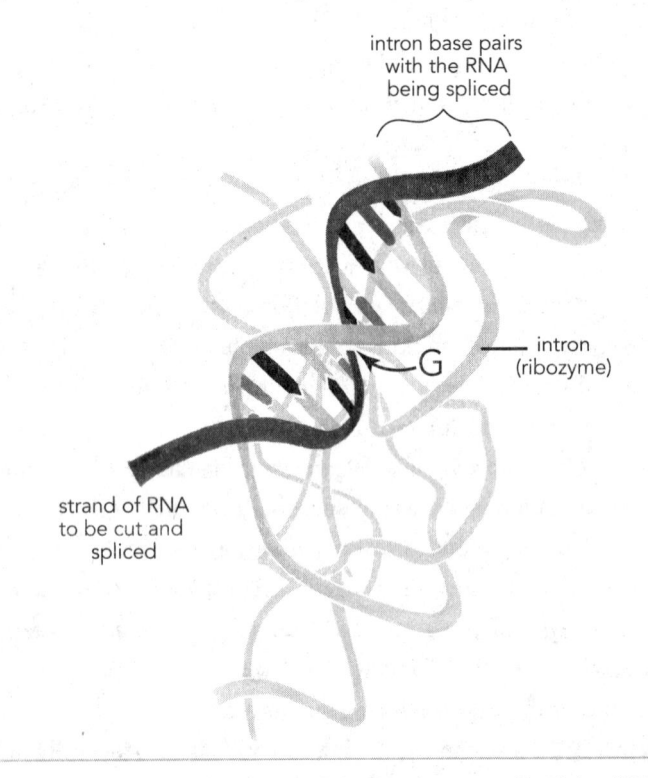

The three-dimensional structure of the *Tetrahymena* self-splicing RNA was modeled by François Michel and Eric Westhof. The intron (light-shaded) pairs with the RNA near the site of splicing (dark-shaded). The intron also binds a molecule of guanosine (G), itself one of the building blocks of RNA, and uses it as chemical scissors to cleave the black RNA strand at the splice site.

like a baby being embraced by its two parents. The "baby" was an RNA helix containing the site that needed to be cleaved and spliced. One of the "parents" was a portion of the RNA structure that François had previously shown to position the required guanosine, which served as "scissors" to cut out the intron. The other parent, called P4-P6 (paired regions 4-6), was supporting the positioning of these key RNA elements.

The model fit together perfectly, but how close was it to the real structure? Finding out would require X-ray crystallography along with the talent and fortitude of a young woman from Hawaii.

LET'S BE CRYSTAL CLEAR

Jennifer Doudna grew up on the verdant eastern shores of the Big Island of Hawaii. Exploring the wonders of the local tide pools and hiking in awe along the rim of the active Kilauea volcano, she got hooked on science at an early age. She crossed the Pacific for college, majoring in biochemistry at Pomona College in California, then the rest of the United States when she moved to Harvard Medical School for her PhD. There, her dissertation research concerned the function of the *Tetrahymena* ribozyme.

This made us more or less competitors, but the rivalry was always friendly. Once, while in the midst of her graduate studies, Jennifer visited me in Boulder, and it was impossible not to be impressed (if not a bit intimidated) by this wisp of a woman. She had an uncanny talent for designing just the right experiment to test any hypothesis, and she possessed more energy and drive than any scientist I'd ever met. Thus, when she obtained her PhD in 1989 and asked to join us in Boulder for her postdoctoral studies, I instantly agreed.

Jennifer and I, along with many others in the RNA community, were convinced that understanding how any ribozyme worked would require a 3D photo of its shape. But RNA structure research was suffering a drought—no large RNA structure had been unlocked since that of tRNA 15 years earlier. We knew that obtaining such a photo was an ambitious undertaking. Yet if we could pull it off it would be a landmark achievement, one that would be destined to show up in textbooks around the world.

A ribozyme structure promised to be a gold mine of answers to fundamental questions about RNA structure more generally, including the infamous mystery of how RNA could fold to give a specific catalytic structure. The ways that proteins fold and form enzymatic active sites were well known. The protein enzyme packs all its greasy side chains on its inside, forming a hydrophobic (water-hating) core, which supports a catalytically active cleft on its hydrophilic (water-loving) exterior. But RNA couldn't possibly use the same principles to form its structure, because it doesn't have any hydrophobic units to play with. Even worse, RNA is negatively charged at every step along the chain, whereas a protein chain is mostly uncharged.* Making a compact structure out of RNA means bringing all those negative charges together, sort of like arranging a lot of magnets with their south poles all pointing inward and then trying to push them together—you get repulsion. The tRNA structure gave only one paradigm for RNA structure, and it wasn't catalytic; it didn't do anything by itself, in the absence of the ribosome and mRNA. The *Tetrahymena* ribozyme structure would give the first picture of how a large RNA could fold like a protein, despite being apparently ill equipped to do so.

Upon her arrival in Boulder in 1991, Jennifer and I agreed that it might be too ambitious to try to solve the structure of the entire *Tetrahymena* ribozyme, which at 414 bases was about six times larger than tRNA. Instead, we decided to aim for a half portion of the molecule as an initial X-ray crystallization target—but not just any half portion; it had to be one that was functionally and structurally worthy of pursuit. Felicia Murphy, a grad student in my lab, had identified a key portion of the ribozyme called "P4-P6" that fit the bill. She found that P4-P6 folded

---

*Of the 20 amino acids found in proteins, 15 are uncharged, 2 are negatively charged, 2 are positively charged, and one—histidine—has the very useful property of switching from uncharged to positively charged as its environment becomes more acidic.

over on itself like an old-fashioned wooden clothespin and was critical for positioning the part of the RNA that contained one of the two splice sites—though the atomic-level details were unknown. It would be Jennifer's job to try to reveal them.

Jennifer teamed up with Anne Gooding, a staff scientist in my lab, and soon they were synthesizing the P4-P6 RNA and setting up crystallization drops under a series of salt conditions. Before long they found a recipe that reproducibly gave beautiful sharp-edged crystals. Presumably the RNA molecules were perfectly aligned in straight rows and columns, but initially, these crystals didn't give very good diffraction patterns in the X-ray beam. The radiation was damaging the RNA in the crystals, preventing a clean image.

In 1993, Joan and Tom Steitz, the two Yale professors who were stalwart members of RiboSki, were spending a year on sabbatical in Boulder. They were visiting my research group and that of my colleague Olke Uhlenbeck. Tom Steitz was one of the world's preeminent X-ray crystallographers, and his group at Yale had solved important structures including key RNA-protein and DNA-protein complexes. Tom enjoyed hanging out in our breakroom talking about science, and in one conversation with Jennifer he described how his group had started freezing their crystals to minimize X-ray damage. They used liquid nitrogen to keep liquid propane very cold, and then plunged the crystal into the liquid propane to freeze it very quickly, before ice crystals could form. Jennifer and Anne mastered the technique and were delighted to find that the subsequent X-ray diffraction pattern of the P4-P6 RNA was much improved. It was good enough that we might be able to see where individual atoms were located within the folded RNA structure.

But, as usual, research progress is two steps forward and one step back—and that's on a good week. Our big "one step back" was a technical problem that frustrated us for the remainder of Jennifer's time in Boulder. Computing a structure requires crystals not only of the molecule that you're studying but also of a "heavy-atom derivative" of it—one with a heavy atom nestled into one or several fixed positions in the molecule.

A "heavy atom" is one with a large number of protons, neutrons, and electrons, such as platinum, gold, silver, mercury, selenium, tungsten, or osmium. Only by comparing the diffraction pattern of the original molecule with that of the heavy-atom derivative can you compute the molecule's 3D structure. Heavy atoms that nestled into pockets of proteins were well known, but RNA was a different animal, and the heavy atoms that worked for proteins just didn't do the job for RNA.

So for the moment the P4-P6 structure remained unsolved. Yet after three years in Boulder, Jennifer had such a stellar reputation from her dissertation work at Harvard and her progress on the groundbreaking RNA structure in Boulder that universities were vying to recruit her. She chose Yale—in no small part because of the relationship she'd already formed with Tom and Joan Steitz. She took along a Boulder graduate student, Jamie Cate, who continued to test metal after metal as a heavy-atom derivative. After many metals failed to do the job, Jamie saw that an osmium ion had the right size to substitute for magnesium ion in the folded RNA. A suitable osmium compound had been synthesized by a Stanford chemist, who was just about to retire. In a stroke of good luck, Jamie contacted him just before his laboratory was cleaned out and obtained a gift of what turned out to be the magic metal. The osmium compound indeed substituted for three of the magnesium ions bound at specific sites in the RNA, allowing Jamie and Jennifer finally to solve the P4-P6 RNA structure in 1996.

The structure was breathtaking. It revealed how the RNA molecule could fold back on itself to form a compact internal core, something that proteins commonly did but which had seemed so challenging for RNA. Yet, for a catalytic RNA, it made sense that it might form a protein-like structure—RNA was acting like a protein, so why shouldn't it also look somewhat like a protein? The structure also showed how positively charged magnesium ions, which are normal constituents of living cells, positioned themselves to solve the charge-repulsion problem of folding a highly negatively charged RNA. Thinking of magnets again, if you want to bring the negative poles of two magnets together and have them stay together, put a positive magnet pole in between.

The next step was to solve the structure of the entire ribozyme. Barb Golden came to Boulder for her postdoctoral research, and in 1998 she achieved a new size record for RNA crystallography: a 247-nucleotide version of the *Tetrahymena* intron that was active as a biocatalyst. This RNA included the P4-P6 domain and, as we'd predicted, the P4-P6 domain in the context of the active ribozyme and the P4-P6 domain in isolation looked almost identical. Barb's intron structure also showed a "cradle" formed by the RNA, waiting to embrace the RNA helix that contained the site of splicing; it looked satisfyingly similar to what had been predicted by Michel and Westhof eight years earlier.

Cracking the structure of the *Tetrahymena* P4-P6 domain stimulated the small field of RNA structure, and soon crystal structures of other large, functional RNA molecules were solved. Taken as a group, these structures provided glimpses of the vast capabilities of RNA. Each functional RNA would of course have its own form, but the general principles seen with the *Tetrahymena* ribozyme seemed destined to be revisited in many other RNAs. The structure was a glorious advertisement for what sort of complex machines could be formed from only A, G, C, and U.

And there, more or less, the field of RNA structure stood for the next decade. Every year, another structure or two would be solved. Scientists were making steady progress; but whereas tens of thousands of protein structures had been solved, less than 1 percent as many RNA-only structures were reported. This was not a happy situation, not only because it slowed our understanding of the fundamentals of RNA but also because it impeded the progress of potentially lifesaving medical innovations. The industrial scientists developing drugs to fight diseases need a detailed picture of their target molecule to guide their searches. For protein targets, they can usually look up the structure, already solved by someone else—or, increasingly, they can calculate it on the basis of the vast database of closely related structures. But the structure of most RNA targets was unknown, inhibiting attempts at drug development. What was needed was a paradigm shift in determining the structures of RNAs.

THE WISDOM OF CROWDS

There's more than one way to predict RNA structure, avoiding the slow and uncertain process of X-ray crystallography. We've seen one approach: find some really smart people with deep knowledge of the principles of RNA folding, such as François Michel and Eric Westhof, and give them several years to solve the problem. But what about the opposite tack—finding thousands of nonscientists who have no experience with RNA structure and letting them each spend a few hours working on the problem? We've all heard about crowdsourcing. Could it work on a topic as arcane as RNA folding, and how many people would even be interested in joining up?

In January 2017, I found myself sitting in Rhiju Das's office at Stanford University with my jaw perpetually dropped. It's always humbling when a fellow scientist tells you some fantastic thing that he or she has done, which you never could have imagined. Rhiju had 37,000 people from all over the world playing a computer game called eterna—or eteRNA—and these gamers were finding solutions to the RNA-folding problem.

In 2009, eteRNA announced its first challenge online: design an RNA that will fold to look like a five-pointed star or to look like a cross. In other words, what sequence of A's, G's, C's, and U's will form the correct base pairs to fold into the target shape? Players came from all walks of life. Some were graduate students doing RNA research, while others were Sudoku aficionados who'd hardly heard of RNA but were eager to try a new kind of puzzle. Some had developed computer programs to fold RNA; others stuck to paper and pencil. Players submitted their answers to the eteRNA website, and then everyone voted for the sequences that they thought were most likely to fold into the target shape and not other shapes. The eight sequences that garnered the highest number of votes were then actually synthesized at Stanford. Each one was tested by a clever method called SHAPE, which was invented by Kevin Weeks, a former postdoc in my lab now on the faculty of the University of North Carolina at Chapel Hill. SHAPE involves treating the RNA with a chemical compound that reacts only with single-stranded nucleotides, then identifying which nucleotides are reactive. For example, if an RNA sequence really folds into a five-pointed star, it should have a characteristic SHAPE reactivity pattern as shown below.

cross bulged cross star

SHAPE reactivity ↑

Winners of the eteRNA contest devised RNA sequences that they predicted would fold into a cross, a bulged cross, or a star. To test whether the predictions were correct, the SHAPE chemical reaction was used to identify the single-stranded regions of each structure, some of which are shown by the arrows.

Thirty-seven thousand players rose to the challenge, and the top ones nailed the solution to each problem. The results were striking enough that they led to a publication that, most unusually, had 100 eteRNA game-players as coauthors. As of 2022, 4,181,632 RNA structure puzzles had been solved by the community, mostly by players who knew little about RNA before they got hooked on the game.

In recent years, eteRNA has raised the stakes and tackled real research problems. For example, in their 2020 Open Vaccine competition, players competed to design an improved version of a Covid-19 mRNA vaccine that didn't require ultracold storage. The hypothesis—or, more accurately, the educated guess—was that designing an mRNA to fold into a highly base-paired structure while retaining the coding capacity for the coronavirus Spike protein would make the mRNA more stable during storage and also when injected into human arms. Because many of the amino acids are specified by two, four, or even six codons, there is an astounding number of sequences that encode the Spike protein. This number—10^{630}— might as well be infinity, because no computer could sort through that many sequences. So eteRNA crowdsourced the problem to any videogamer who wanted to play and waited for the answers to come in. Because the contest was launched on March 18, 2020, when many people were locked down at home because of Covid-19, there were many takers.

The gamers came up with many mRNA sequences that were real "superfolders," with most of the bases locked up in base pairs. But were they in fact more stable? At Stanford, eight of the superfolding RNAs were tested for their stability both during storage and also once put into human cells. Gratifyingly, these sequences were up to two times more stable than those designed by current computer programs. Now the Stanford researchers were holding their breath: were these RNAs so tightly folded that they wouldn't thread through the ribosome to make the Spike protein? Not to worry, *translation* from mRNA to protein worked just fine. Finally, they turned their superfolder mRNAs over to the Pfizer vaccine group, which tested their longevity in a vaccine formulation. After sitting at warm temperatures for two weeks, the superfolder vaccines were mostly intact, faring much better than mRNA vaccines designed with current technology. Because the approved Covid-19 mRNA vaccines need to be stored and shipped at ultracold temperatures, which makes it challenging to deliver them to poor countries, this increased thermal stability has exciting potential to one day become the basis of a more accessible and cheaper vaccine.

ARTIFICIAL INTELLIGENCE TO THE RESCUE?

Harnessing the wisdom of the crowd is a unique way of solving RNA structures. But the future of the field will more likely rely on replacing human brainpower with machine learning. Artificial intelligence (AI) can already write newspaper articles and social-media posts; it can translate spoken sentences into text; and it can—in principle, anyway—allow self-driving cars to get around town safely. So can we turn it loose to predict RNA structures? Could it have saved Michel and Westhof the seven years of work it took them to predict the structure of a self-splicing intron? Could it have saved my lab the seven years we spent solving the structure of a self-splicing intron by X-ray crystallography? Could it save 37,000 people from playing eteRNA to find the best sequences to fold into a star or a cross? The answer is almost certainly "yes," and although we're not there yet, the future seems clear.

In 2021, the eteRNA cocreator Rhiju Das and his Stanford colleague Ron Dror announced a breakthrough. They managed to design an AI computer program that can predict with considerable success what 3D structure a given RNA sequence will adopt when it folds up. One challenge they faced was assembling an adequate "training set." AI programs need to train on some real information before they can set off into the unknown. AI has no trouble sorting photos of dogs from photos of cats, because the program is trained by millions of images on the internet labeled "dog" or "cat." But for RNA folding, Das and Dror had only 18 sequences paired with structures for their training set. Furthermore, they were asking AI to do something much more difficult than just recognize an RNA structure when it saw one; they wanted it to use the sequence of A, G, C, and U nucleotides to predict the correct 3D structure. Remarkably, the 18-sequence training set was enough for their program to outperform previous structure-prediction methods.

How does one measure the success of RNA 3D-structure predictions? Eric Westhof has recruited RNA structural biologists from all over the world to play a bit of a game, which he calls RNA-Puzzles. When a participating scientist solves a new RNA structure, for example, by X-ray crystallography, that scientist agrees to hold off making the solution publicly available until the players have had a month to take their best shots at predicting the structure on the basis of the nucleotide sequence alone. In short, the players are blinded to the answer. Some players have developed automated webservers for RNA structure prediction, while others take a more build-it-by-hand approach. All the entries are gathered, and when the deadline arrives, the true structure is revealed.

For four out of four RNA-Puzzles, the Das-Dror AI method gave the most accurate model submitted by any participant. It's not yet perfect, but it's well on its way to being a reliable tool for the community. One can see a time in the future when such structures will be solved without anyone going into a laboratory—an exciting prospect for scientific progress, but a rather sad prospect for those of us who've spent our lives cooking up molecular recipes in the lab and savoring the results.

Seeing tRNA and then ribozymes in three dimensions had a huge impact on RNA science. Scientists could see how the tRNAs folded so they could slot into the ribosome, where they not only brought in the correct amino acids but also aimed the amino acids at each other to encourage them to react and build a protein chain. Scientists could see the atomic details of how the *Tetrahymena* ribozyme positioned guanosine to attack at an RNA splice site, how RNase P catalyzed the cleavage of a specific bond to create a mature tRNA, and how snRNAs orchestrated mRNA splicing.

Beyond simply trying to understand how biology works, having an RNA structure allows scientists and bioengineers to design variants of the structure to repurpose it for new applications. For example, a bioengineer might design a ribozyme that can be used as part of a molecular circuit to detect a toxic compound or a particular virus in an environmental sample. There's an entire field of synthetic biology that now uses ribozymes as sensors and switches, but they benefit from knowing the RNA structure as a starting point.

The breakthroughs in determining 3D structures of larger and larger RNAs encouraged people in the field to ramp up their aspirations. A few intrepid scientists even started setting their sights on the cell's final frontier, the mother of all molecular machines, whose secret power source had long been a mystery. They aimed to solve the structure of the ribosome.

Chapter 5

THE MOTHERSHIP

Harry Noller isn't like most other biochemists. Rarely do the words "cool" and "biochemist" appear in the same sentence, but Harry is cool. A professor at the University of California, Santa Cruz, he's also a jazz musician who has played saxophone with Chet Baker. He spends his spare time refurbishing vintage Ferraris. He's also a natural-born linguist. I've attended overseas conferences with Harry, and regardless of where we are in the world, when we sit down at a café, he seems able to order in the local language.

The University of California, Santa Cruz, has an eye-popping setting, nestled in towering redwoods at the north edge of Monterey Bay. When Harry set up his research laboratory there in 1968, his goal was to understand how the ribosome worked. This powerful molecular machine, which makes all the proteins in all living things, is a true wonder of nature. Like a locomotive on a railroad track, the ribosome rides along a messenger RNA. It stops for an instant at each triplet codon, waits for the correct transfer RNA to pair with it, then adds the correct amino acid to a growing protein chain. And it is impressively versatile: give it a thousand different mRNAs, and it produces the thousand corresponding proteins.

When Harry began his research, everyone still assumed that proteins were the only things in nature that could catalyze biological reactions, so Harry made it his mission to figure out which proteins in the ribosome did the hard work of protein synthesis. Hypothetically, one ribosomal protein might bind the mRNA, one or two might bind the tRNAs, and one might catalyze what scientists call *peptidyl transfer*—the chemical reaction that joins two amino acids together.

Never mind that the ribosome was only one-third protein by mass, and the other two-thirds was made up of ribosomal RNAs. Scientists believed those RNAs—there were three in bacterial ribosomes, including the *E. coli* ribosome that Harry studied—must provide some sort of scaffold that helped the key proteins organize themselves. In other words, protein was king, and the ribosomal RNAs were just dim-witted peasants serving the king.

But things didn't go so well for the "let's find the key proteins" plan. Harry set up a system where he could build up ribosomes from their constituent parts—RNA and proteins—and then see if those reconstituted ribosomes were active in protein synthesis. This allowed him to leave out one protein at a time and ascertain which ones were essential—sort of like baking bread and leaving out one ingredient at a time to see which ingredients were essential. In the case of the ribosome, Harry left out one protein at a time, and pretty much nothing happened—the ribosome continued to do its job. Disappointing, and quite perplexing. Where were those key catalytic proteins?

In 1972, an undergraduate student in Harry's lab, Jonathan Chaires, needed to finish up his senior thesis, so Harry suggested they try a radical new approach. Harry knew that a chemical called kethoxal reacted very specifically with G bases in RNA, making them a few atoms larger than normal, without affecting the neighboring proteins. Harry had been getting nowhere by futzing with the ribosome's proteins. Maybe the key was to leave them alone and play around with the RNA?

Their test for protein synthesis would be the same one Marshall

Nirenberg had used to help crack the genetic code: take *E. coli* ribosomes (either kethoxal-treated or untreated), add poly(U) as a synthetic mRNA, and look for production of the chain of amino acids, polyphenylalanine. The first time Harry and Jonathan treated the ribosomes with kethoxal, they saw that it stopped protein synthesis dead in its tracks. What's more, only 10 out of the hundreds of G's in each ribosomal RNA had reacted with kethoxal, and that had been enough to derail protein synthesis. The ribosome really didn't like having its RNA messed with.

On the basis of this experiment, it appeared to be the ribosomal RNA, not one of the ribosomal proteins, that did the key job of binding tRNAs. Harry felt like his Ferrari had just plunged off California's Highway 1 into the Pacific. Where to go from here?

COURSE CORRECTION

To unravel the ribosome, Harry would have to become an RNA guy. This is a recurring event in science, when a scientist has been working away to prove one hypothesis and the data suddenly indicate that the truth might lie in an entirely different direction. It was very similar to what I experienced 10 years later, when my lab was chasing the furtive protein enzyme that *must* be hiding in our RNA splicing reactions, only to come to the realization that the RNA was splicing itself. Deciding which path to take at these crossroads is never completely clear, and many scientists are so focused on their training in a particular field that they are reluctant to take a leap. As the saying (often attributed to Winston Churchill) goes, "Men occasionally stumble over the truth, but most of them pick themselves up and hurry off as if nothing had happened."

Not Harry Noller. He knew that to understand its function, he would need to understand the structure of the ribosomal RNA. But RNA structure was largely inscrutable in 1972, with the tRNA cloverleaf the lonely exception. Harry knew that the larger the RNA was, the more difficult it would be to figure out its structure—which was bad news for him, because two of the three ribosomal RNAs in all species were gigantic: in *E. coli*, one weighed in at 1,542 nucleotides and the other at 2,904 nucleotides. The third one was smaller, 120 nucleotides, but still larger than tRNA.

Harry was on sabbatical in 1975 when an epiphany hit. With extra time to spend in the library, he came across a recent paper by Carl Woese, a University of Illinois microbiologist. A few years later, Carl would discover an entirely new domain of life, the Archaea, organisms that inhabit the sulfurous hot springs of Yellowstone National Park, among other seemingly inhospitable environments. But back in 1975, Carl and his research associate George Fox had nailed the 2D structure of the smallest ribosomal RNA, consisting of 120 nucleotides. Their method was based on the same idea that had worked for the tRNA cloverleaf structure in the 1960s. They had the sequences of this small ribosomal RNA from a dozen different organisms, mostly bacteria and one frog. Form follows function, so they assumed that these RNAs would all fold up in the same way, despite their species-specific sequence differences, because these RNAs were all presumably playing the same role in the ribosome. Of the many ways in which the RNA structure could be folded, respecting the rule of A-U and G-C base pairs, only one of the shapes worked for *all* the dozen organisms. A light went on in Harry's head: this would be the road to the structure of the large ribosomal RNAs, even if it would be uphill and take years.

When Harry phoned Carl Woese in 1975, each found a kindred spirit in the other. They were the two outliers. Ninety-nine percent of scientists still thought that it must be the proteins inherent in the ribosome that were doing the heavy lifting and using mRNA to make new proteins. The other 1 percent, who believed it was the ribosomal RNA doing the heavy lifting, comprised Harry and Carl. "They didn't take us seriously," said Harry. "But the upside was that we had a decade with zero competition."

Because Santa Cruz, California, and Urbana, Illinois, are not in convenient proximity, the collaboration proceeded mostly by telephone and by mail, trading lists of the sequences of short scraps of ribosomal

RNA. These short scraps were obtained by chopping up the RNA with an enzyme called ribonuclease T1, which cuts after every G in the RNA alphabet. The process was like running a page of a document through a paper shredder. This was necessary because, at that time, only short pieces of RNA could be sequenced. In the 1,542-nucleotide rRNA, there were more than 100 "words"—short strings of RNA nucleotides. The Woese and Noller teams figured out how these words were spelled (e.g., CUCAG and UACACACCG). After they had spelled all the words making up the rRNA, they next had to assemble them into one long sentence. This was challenging, as difficult as taking the output of a paper shredder and reassembling the original document. Only then were they ready to announce the complete sequence of the 1,542-nucleotide RNA.* Once they had the RNA sequence in hand, they could see how parts of the sequence would pair with each other to fold up the RNA—just as Fox and Woese had done for the smallest ribosomal RNA and as François Michel would later do for the ribozymes.

The resulting 2D map of this rRNA was announced in 1980. It looked somewhat like the terminal map of O'Hare International Airport in Chicago: many concourses protruding from a central hub, some of them branched like the letter Y.

It was followed a year later by the 2D structure of the 2,904-nucleotide ribosomal RNA, an even larger set of terminals and concourses.

These two maps of what Harry and Carl called "the mothership" would give hundreds of ribosome biologists around the world a framework to plan and interpret their experimental results. But, ultimately, maps were not enough. Catalysis does not take place in two dimensions, and there was no way to truly understand how the mothership worked

---

*Remarkably, sequencing this RNA—which required about 20 person-years to complete in the 1970s—is now routinely done in a single day by automated sequencers. This technical advance has enabled the human microbiome projects, because the ribosomal RNA sequences immediately identify what bacteria are present in an environmental sample or in a swab from a location on a human body.

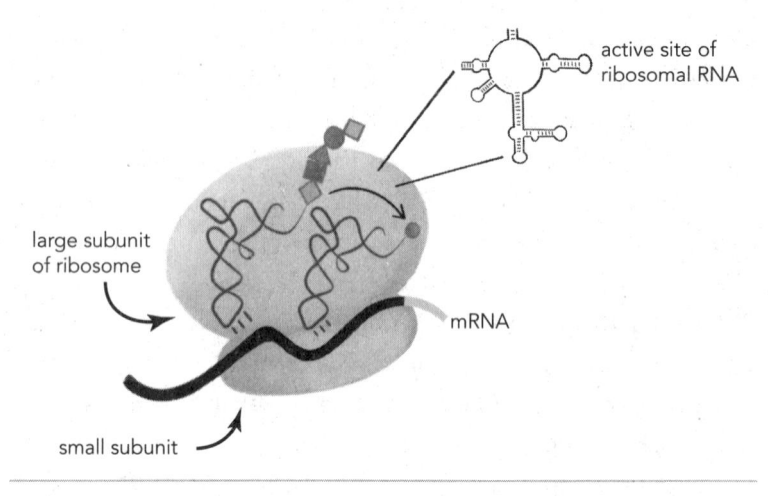

The RNA of the large subunit of the ribosome folds into a multibranched structure; only the portion that catalyzes the peptidyl transfer reaction is shown here (structure at upper right). The ribosome illustrated contains two tRNAs, bound to adjacent triplet codons on the mRNA. One tRNA carries the growing protein chain (shapes represent five different amino acids), while the next amino acid to be added to the chain (single dark sphere) has been brought into the ribosome by a different tRNA. Because of the difficulty of portraying the 3D object in a 2D diagram, the amino acid–carrying ends of the two tRNAs appear separated here, but they are actually held close together in the 3D ribosome.

without seeing it in real life. And, in a pattern that recurs time and again, figuring out how the ribosome worked not only advanced the cause of science but also laid the foundation for potentially lifesaving improvements to antibiotics—improvements that Harry could not have dreamed of when he began his quest to unravel the ribosome.

HIBERNATING LIZARDS AND THE DEAD SEA

Harry Noller may have been almost alone in his conviction that the ribosome used RNA as its secret power source, but he was not alone in wanting a clear picture of it. Bioscientists around the world craved a three-dimensional view of the molecular machine that synthesized the proteins in every organism. But getting a sharp picture turned out to be as elusive as photographing the Loch Ness Monster.

As with tRNA and with ribozymes, the technique du jour was X-ray crystallography. Ada Yonath started laying the groundwork for the crystallization of bacterial ribosomes in the 1970s at the Weizmann Institute of Science in Rehovot, Israel. She, like the many other researchers attempting ribosome crystallization around the world, found it incredibly difficult, meeting with failure after failure. She might have given up, but she took comfort in reports that ribosomes packed themselves into crystalline arrays in hibernating bears and in hibernating southern Italian lizards. If ribosomes could form ordered crystalline-like arrays in living animals in the cold, she reasoned, then she should be able to get them to crystallize in the laboratory.

By 1980, Yonath was able to grow crystals of bacterial ribosomes that diffracted X-rays reasonably well. But the bacterial ribosomes were not stable in high-salt solutions, the preferred medium for crystallization; some of the proteins fell off the ribosome, leaving a mixture of incomplete ribosomal particles. So, she and her colleagues reasoned that a salt-loving organism might have ribosomes that remained stable in high-salt conditions. Given their proximity to the Dead Sea, they tried *Halobacterium marismortui* ("salt bacterium Dead Sea"), and eventually its ribosomes turned out to be winners.

Given all this progress, it would seem that cracking the 3D structure of the ribosome—with its three RNA molecules and 55 proteins—might be right around the corner. But, in fact, it took another 15 years. For, as Jennifer Doudna and my lab had encountered when striving to solve the *Tetrahymena* ribozyme domain structure, having good crystals that diffract X-rays is only half the battle. The other half is solving the "heavy-atom problem" once again. In order to calculate an RNA's 3D structure, clear X-ray diffraction data are needed for the RNA molecule with and without a heavy atom bound to it. This is where Tom Steitz and Venki Ramakrishnan come into the story.

UNVEILING THE CRYSTAL PALACE

So far, we've been speaking of the ribosome in the singular. But in fact the ribosome is not a single entity but a pair of enormous complexes each composed of RNA and proteins—called the large and small subunits—that come together to do the job of protein synthesis in all species. The *small subunit of the ribosome* contains the second largest (1,504 nucleotides) of the three types of rRNA and 22 proteins. It is the first to assemble with the mRNA. Then the *large subunit of the ribosome*—consisting of the other two rRNAs (2,904 and 120 nucleotides) and some 33 proteins—makes its entrance. The large subunit houses the catalytic center, which is responsible for stitching together amino acids, one after another, to produce the chain of amino acids that we call a protein. These details are important because the next two actors in our drama each tackled a different subunit: Tom Steitz for the large subunit, and Venki Ramakrishnan for the small.

By 1995, Tom Steitz had an unparalleled track record for solving the structures of the most fundamental molecular machines in biology. He had determined the structures of DNA polymerases, which replicate the parental double helix into two daughters. He had determined the structures of RNA polymerases, which copy the information from DNA to RNA. He had determined the structure of the HIV reverse transcriptase, which copies the HIV RNA into DNA that is then inserted into a human chromosome. And he had determined the structure of an enzyme that adds the correct amino acid to a tRNA molecule.

But could Tom crack the mothership? In 1995, he assembled a team of three postdoctoral fellows who were ready for this adventure. They were joined by Tom's longtime friend and Yale colleague Peter Moore, a ribosome expert. They chose the Dead Sea bacterium as the source of their ribosomal subunits, because Ada Yonath had already paved the way. The Steitz group focused on the large subunit—which catalyzed the joining of amino acids into proteins—and to crack its structure, they'd need to solve the dreaded heavy-atom problem.

A skilled sailor, Tom explained this problem with a seafaring story. He compared the heavy-atom measurements in X-ray crystallography to the problem of measuring the weight of a ship captain by subtracting the weight of an empty boat from the weight of the boat plus captain. If the boat were a small sailboat, this would work reasonably well. But what if the boat were the RMS *Queen Mary*? Then subtracting the weight of the *Queen Mary* from that of the *Queen Mary* plus captain would be a really difficult way to determine how much the captain weighed. And the ribosome, with its 250,000 atoms, was the *Queen Mary* of biomolecular machines.

A key moment came when the Steitz team found a solution to the "how to weigh the ship captain" problem . . . use an extremely heavy ship captain. It was a cluster of 18 tungsten atoms, which just happened to nestle into a specific crevice in the ribosomal subunit. Because tungsten is the element used to make the filaments that glow inside a traditional lightbulb, you might say that this trick lit up the structure. Over the next years, the Steitz lab revealed a succession of better and better photos of the large subunit of the ribosome, culminating in a hitherto unimaginably sharp image in the year 2000.

What's it like to solve the 3D structure of a biomolecular machine? It's as if there's a crystal palace that's been covered with a huge shroud for a long time. Hundreds of researchers have used indirect methods to glean what's inside—there must be a kitchen, a dining room, and multiple bedrooms and bathrooms. But no one knows how the rooms are arranged relative to each other. What's the layout, and how does the floor plan support the palace's functions? And then, in a moment, the structure is solved. The huge shroud is pulled off, and one can peer through the transparent walls to see everything within, and one can even walk through all the rooms. That's what it's like to solve an atomic structure by X-ray crystallography. You suddenly see the details of everything that scientists have proposed, and probed, and posited in laboratories all over the world for many years, and you see which of the ideas are right and which are wrong.

For the Steitz team, a moment of revelation came when they peered into the catalytic center of the large ribosomal subunit and saw that it was made exclusively of RNA. There were no proteins in the vicinity.

Everything that Harry Noller and Carl Woese had proposed about the centrality of the RNA on the basis of their painstaking experiments and well-aimed intuition proved to be true. The ribosome is in fact a ribozyme, a catalytic RNA machine. Certainly the RNA is supported by a cast of proteins, much as the RNase P RNA is supported by a protein that helps keep the RNA well organized under cellular conditions. But the heart of the ribosome is pure RNA.

But at this point, our story is only half told. Yes, the structure of the large subunit of the ribosome showed in atomic detail that it was an RNA enzyme, not a protein enzyme, that catalyzed the stringing together of amino acids into proteins. But how about the key steps required for reading out the code in the mRNA and lining up the proper tRNAs, the adaptors that would determine which amino acids were strung together? The secrets of decoding the message would lie in the small subunit of the ribosome.

Venki Ramakrishnan grew up in India, received his PhD in physics at Ohio University, and then got hooked on ribosomes as a postdoctoral fellow at Yale. In 1995, Venki joined the faculty of the University of Utah and turned his attention to solving the structure of the ribosome's small subunit. He and his graduate student Bil Clemons perfected methods to grow decent crystals of the ribosome subunits and then came face to face with the dreaded heavy-atom problem. They soaked in every heavy atom they could obtain, and as with the large subunit in the Steitz lab, it was clusters of tungsten atoms that finally lit up the structure. In 1999, Venki moved to the vaunted Laboratory of Molecular Biology in Cambridge, England, and within about a year his team finished the work that they'd begun in Utah. They solved the structure of the ribosome small subunit. They were peering into their own crystal palace, seeing it in astounding detail.

But something was missing from the small-subunit structure. The family that lived in the palace, the mRNA and the tRNAs, were not home, for the simple reason that they had not been included in the crystallization mixture. The purpose of studying the ribosome was to understand protein synthesis, and, as we've already learned, the ribosome doesn't make protein all by itself. It needs mRNA to specify which protein gets made, and it needs tRNAs to bring in the matching amino acids. So it was challenging just looking at the ribosome structure to understand how proteins got built. You needed to see where exactly the mRNA and tRNA fit in; you needed to see the house and its occupants.

The challenge to visualize the entire ribosome inhabited by its functional family members, the tRNAs and mRNA, fell to Jamie Cate, who had moved to Harry Noller's lab at UC Santa Cruz for postdoctoral research. Luckily, he was well prepared for this job, having cut his teeth at Yale with Jennifer Doudna pinning down the structure of the ribozyme. In 1999, Jamie and Harry succeeded in cracking the first-ever crystal structure of a ribosome in its functional state, inhabited by tRNAs and mRNA. However, their crystals diffracted X-rays to a limited extent, so the resulting photograph was a bit blurry. They were looking at their crystal palace through foggy goggles.

But the two less-than-perfect views—the somewhat cloudy image of the entire ribosome with all its partners from Jamie and Harry and the very sharp image of a vacant small subunit from Venki—complemented each other brilliantly. Superimposing the locations of the mRNA and tRNAs on the high-resolution structure revealed how the ribosomal RNA bases were helping read out the mRNA code. Some rRNA bases were holding the tRNAs in place, while others were positioning the mRNA to decode it.

Everything revolved around RNA. The functional sites that held the tRNAs and the mRNA consisted almost entirely of RNA. The critical surface that connected the small ribosomal subunit with the large one? Again, mostly RNA. Only one of the small subunit's 22 proteins was helping out. All the rest of the action was clearly being organized by RNA.

In 40 years, science had gone from deciphering the mRNA code to seeing in astonishing detail how it was decoded to achieve protein synthesis. Once again, Noller and Woese were vindicated. They could no longer bemoan being in the 1 percent minority who believed that RNA was the key to protein synthesis, with the proteins playing bit parts. Seeing is believing. Now the whole scientific world was forced to face up to the fact that RNA was king.

WHO'S DRUGGING WHOM?

That the protein-synthesizing ribosome runs almost entirely on the power of RNA might be a mind-blowing revelation for biochemists such as Harry and me, but we'd forgive you for wondering why anyone else should care. Are there any practical benefits to understanding the structure and function of ribosomal RNA?

Consider antibiotics. Unlocking the ribosome structures has given us previously unimaginable insights into how many of the antibiotics work, how antibiotic resistance can arise, and how antibiotics can be improved in the future.

An effective antibiotic needs to disrupt a vital bacterial process without affecting related human processes. You might think that the ribosome would be a bad target, because its fundamental features—large subunit, small subunit, binding tRNAs and mRNA, catalyzing the assembly of amino acids—appear in all life-forms. But it turns out that in the billion years since humans and bacteria have been going their own ways, evolutionarily speaking, the human and bacterial ribosomes have diverged from each other just enough that you can find drugs that inhibit only the bacterial ribosomes. Remarkably, about half of all useful antibiotics target bacterial ribosomes.

Antibiotics were just coming into widespread medical use in the 1960s, spawning huge interest in understanding how they worked. This was the same time that ribosomes, mRNA, tRNA, and the genetic code were being discovered, and the two fields converged. It turned out that many common antibiotics—including those that treat tuberculosis, gonorrhea, and even acne—killed bacteria by inhibiting their ability to make proteins. Scientists quickly found that these antibiotics were binding directly to bacterial ribosomes.

A sense of scale is useful here. A typical antibiotic drug molecule is made up of perhaps 100 atoms, while a bacterial ribosome has about 250,000. The ribosome is 2,500 times larger than the drug. Like the proverbial monkey wrench thrown into a machine, a small drug can inactivate a large ribosome if the drug binds to a functionally critical site within the ribosome. If a drug bound only to the ribosome's outside surface, it would do no harm, and it never would have made the cut to be an antibiotic. For this reason, seeing the antibiotics bound to a bacterial ribosome is of interest not only to the pharmaceutical industry but also to basic scientists who want to understand how the ribosome works.

One early key to that understanding was antibiotic-resistant bacteria. Then, as now, as soon as any antibiotic starts to be widely used, some lucky bug will happen to have a mutation that protects it from the antibiotic. As its neighbors pass away, that lucky bug will multiply and take over the population. It seems inevitable: as soon as there's an antibiotic that effectively kills some type of bacterium, antibiotic resistance will pop up. How can a mutation in the ribosome confer antibiotic resistance?

Think again of the ribosome as a locomotive moving along a messenger RNA railroad track. Each antibiotic is like a monkey wrench of a very specific size and shape. One wrench might fit into a piston of the engine and prevent it from moving, while another might slide into a driving wheel and stop it from turning. There are a hundred ways various wrenches could muck up a locomotive, just as there are a hundred ways various antibiotics can muck up a ribosome. Now imagine that one locomotive has some subtle design differences from the others. Its pistons are a different size, and the slot that leads to its driving wheels is a bit narrower. The wrenches that mucked up other locomotives now have no effect on our new locomotive; it is resistant.

Because antibiotic resistance is so common, scientists had no trouble getting their hands on various antibiotic-resistant ribosomes. In each case, they asked the question: Where is the mutation that confers antibiotic resistance? Is it in the ribosomal RNA or in one of the many ribosomal proteins? Starting in the 1970s, scientists sequenced the ribosomal RNA and ribosomal proteins from the drug-resistant cells and found examples of both situations. In some cases, an amino acid sequence change in a ribosomal protein conferred antibiotic resistance. But in other cases, it was a base sequence change in one of the ribosomal RNAs that did the trick. Those latter cases gave early encouragement to Harry Noller, Carl Woese, and their colleagues in the RNA research community, supporting the still nascent idea that the ribosomal RNA was critical for ribosome function.

All this evidence was a bit indirect, so when ribosome X-ray crystallography fell into place around the year 2000, a number of researchers including Tom Steitz, Venki Ramakrishnan, and Ada Yonath jumped at the chance to see exactly where the antibiotic drugs were sitting in the ribosome. Ideally, they would have wanted to look at ribosomes from the pathogenic bacteria that were the targets of the antibiotics. But those ribosomes hadn't been crystallized, so they added the antibiotics to ribosomes from related bacteria, reasoning that they should work similarly, and took a look.

The Steitz lab captured photos of seven different antibiotic drugs bound to the large ribosomal subunit, and they were all bound to its catalytic center. Each drug bound in a slightly different position, but in all cases the binding would clearly prevent the ends of the tRNAs from docking into the ribosome to engage in protein synthesis. Furthermore, each drug bound to the large ribosomal RNA, not to a protein. After all, if you're going to win a chess match against the ribosome, it may be better to take out the RNA king rather than one of the protein pawns.

Erythromycin, which is effective at treating bacterial infections such as strep throat, was found to bind to a specific site on the large ribosomal subunit. During protein synthesis, the ribosome extrudes whatever protein it's building out of an "exit tunnel." Erythromycin was found to bind in a position that blocked that exit tunnel, thereby preventing the extension of the growing protein chain. Tom Steitz enjoyed referring to this mechanism of inhibition as "molecular constipation."

The small subunit of the ribosome had its own vulnerabilities that antibiotics could exploit. Venki's group snapped photos of six antibiotics, including streptomycin and tetracycline, stuck to the RNA of the small ribosomal subunit. Each of these gave some insights about how the ribosome works. Let's consider spectinomycin, which is used to treat gonorrhea. One of the ribosome's tricks is *translocation*, the movement of an mRNA codon with its bound tRNA from one site to another within the ribosome. This needs to happen each time a codon is read out, in order to make room for the next tRNA to enter.

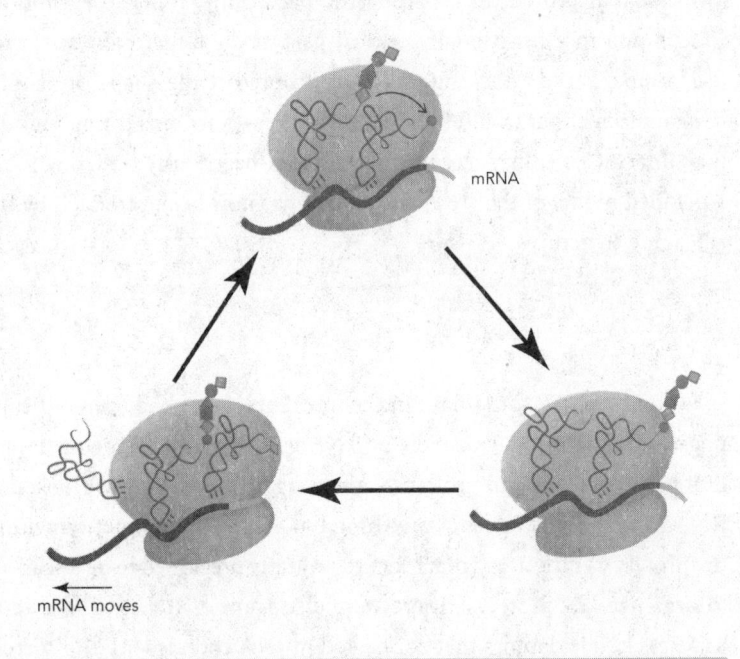

The ribosome uses the information in the mRNA codons to connect amino acids into a protein chain. The peptidyl transfer reaction, shown by the curved arrow (top), results in a chain that is one amino acid longer (bottom right). The mRNA then translocates, moving the two tRNAs to new sites in the ribosome and making space for the next tRNA (carrying the diamond-shaped amino acid) to bind (bottom left). This cycle is repeated for each amino acid that's added.

This translocation step requires the movement of the "head" of the small subunit. It's like nodding your head—one nod per translocation step. Spectinomycin is a rigid molecule, literally a little wrench, and it slots into a specific niche in the ribosomal RNA near the pivot point of the head. This blocks the head from nodding, preventing translocation. No wonder spectinomycin kills bacteria: if their ribosomes can't nod, they can't make any of the proteins they require to live.

The various ribosome structures promise to give the biomedical community a powerful new tool for fighting antibiotic-resistant bacteria. In what's called structure-based drug design, scientists examine the surface of a disease-causing protein or, in the case of antibiotics, any essential protein in a pathogenic bacterium. When they find an indentation in a functionally critical part of the target molecule, they use computerized "docking" software to predict the shape of a small drug molecule that would fill the indentation—a wrench fitting into the machinery. Critically, one can't do structure-based drug design without a detailed model of the structure, which has now been provided by the ribosome structures.

In 20-odd years, RNA's image in the world of science had been radically upgraded. The mid-1960s view of RNA gave it credit for being a conduit between DNA and proteins—a message. Then there was ribosomal RNA, which didn't code for anything but served some function within the protein-synthesizing machinery—initially presumed to be a scaffold to organize key proteins. There were also transfer RNAs, understood to be essential adaptors that connected mRNA codons with the correct amino acids but not considered harbingers of a vast array of noncoding functions for RNA. Then in the 1980s came the discoveries that RNA could be a biocatalyst, that snRNAs were orchestrating mRNA splicing, and that ribosomal RNAs were directly responsible for one of the most central processes in all of life—protein synthesis. RNA had been transformed from backup singer to a star on center stage.

These were all seismic developments that rewrote biology textbooks and would lead to better understanding and treatment of human disease. But RNA research was not just rewriting the rules of science in the here and now. It was also about to shed light on one of our oldest and most profound questions: How did life on our planet begin?

Chapter 6

ORIGINS

The first time I saw Mesa Verde in southwestern Colorado was more than 50 years ago. I don't recall all the details, but I do still remember the early-morning chill on my skin as I climbed the wooden ladder that led up the cliffside.

"Watch your step," warned the National Park ranger. "There's still frost on the rungs."

I also remember the view when I made it to the top, how the full splendor of the ancient village's sandstone walls emerged, glowing gold in the morning sun. The 150 rooms, tiny by today's standards, were perched improbably in an enormous alcove in the cliff. Though this site had been abandoned nearly a millennium ago, the stone towers and cliff dwellings looked remarkably well preserved.

When the rest of our group ascended, the ranger motioned us to gather above a round stone-lined room that appeared to be sunk into the rock.

"The Ancestral Puebloans built these kivas for their spiritual ceremonies," she explained. "When you look down, you'll see a large round structure in the floor—that's the firepit. Now, see the small round hole in the floor? That's the sipapu. The Pueblo people believed that's where humans first came into this world."

I already knew, the first time I saw Cliff Palace, that every culture had its own creation story, whether it was God's busy week in Genesis, or Gaia emerging from a flash of light, or the Pueblo ancestors crawling up through a sipapu. The question of how we got here, how life began on Earth, is perhaps *the* fundamental question. But as much as I've spent my career studying the constituent parts of life, the molecular underpinnings of creation, I thought for many years that the question of how it all began was better suited to philosophers and theologians than chemists like me.

That isn't to say one can't approach such a question chemically. Another way of asking how life began on Earth is asking how the inorganic morphed into the organic. Long, long ago, there was no living thing on our planet, not even the most primitive life-form, only rock and ocean. Then, a moment later, there was life. What did that primordial living thing look like? How did it arise?

Any conversation about the origin of life needs to start with a definition of what that four-letter word actually means. Scientists are not of one mind on this issue. It's been said that there may be as many definitions of life as there are people trying to define it. Many definitions require an entity to grow, to undergo metabolism, and to respond to stimuli. Yet the simplest definitions of life have just two basic requirements: a living thing must be able reproduce itself, and it must be able to mutate.

The first of these requirements—reproduction or replication—seems obvious; it is essential for a living entity to perpetuate itself into future generations. This distinguishes it from a nonliving thing, such as a rock, which does not reproduce. You can stare at a rock for a million years, and you won't see any rock offspring appear. The second requirement, mutation, may come as a surprise. After all, isn't mutation a bad thing? Mutation means making occasional mistakes while copying the information handed down to the next generation—in the case of nucleic acid replication, the four letters of the DNA or RNA alphabet are copied with high but not perfect fidelity. If replication were perfect, then primitive life would remain primitive. No alternative forms would arise; and these variant forms are necessary for natural selection. Mutation is needed to give the descendants of a life-form a chance to improve over generations, to adapt, to evolve.

So, for the purposes of our discussion, let's rephrase the origin-of-life question: "How did the first entity capable of replication and evolution arise?"

When scientists have pondered the origin of life, they have immediately encountered a problem. If life means replication, then there must be some instructions—some information—that get passed down from one generation to the next. In modern life-forms, that information is found in the double helix of DNA. But while DNA gives us the instruction manual for life, it can't copy itself without outside help. Little protein machines called *replicases* act like molecular Xerox machines, copying each parental strand of DNA into a daughter strand, converting one double helix into two copies.

This explains why the origin of life is often seen as the mother of all chicken-or-egg problems. Scientists could never figure out which came first, the informational molecule or the functional molecule, DNA or the protein that reproduces it. Both things essentially had to occur simultaneously—yet the idea that random chemical reactions could have produced DNA *and* its protein-powered copy machine at the same time and in exactly the same place seemed inconceivable. Equally implausible was the possibility that one of these essential elements evolved first and then simply waited a few million years for the other to spring into existence. These substances have limited stability, dictated by the laws of chemistry, and they would have disappeared if they didn't replicate.

So, to solve the origin-of-life conundrum, scientists needed somehow to find a molecule that could play both roles—carry the information, the code for life, *and* reproduce that code all by itself. In other words, we needed the chicken and the egg to be one and the same.

IT'S A SMALL (RNA) WORLD

After my research group discovered RNA self-splicing in *Tetrahymena*, the phone started ringing with invitations from other universities to speak about our ribozyme work. These are important opportunities, especially for an early-career professor. Each time you showcase your research, the audience may include faculty who will later review your grant applications or your submitted manuscripts, as well as graduate students who could be stimulated to apply to your lab for postdoctoral research. Furthermore, highlighting your trainees' work to a big audience helps them get good job offers. In other words, as in many other careers, networking is key.

So I tried to accept as many invitations as I could fit in. In the 12 months after publishing our key paper about the ribozyme in 1982, I crisscrossed the country, giving lectures at a dozen universities and five conferences. It was rather exhausting, but I was getting our research out there, receiving valuable feedback, and meeting new people. I felt like I was headed in the right direction, but I had no idea that I was about to fall headfirst into the dark hole of a sipapu.

In November 1983, I arrived in Los Angeles to give an evening talk at UCLA, expecting it to be just one more research seminar. The invitation had come from something called the Evolutionary Group, which didn't surprise me, because biochemists talk about evolution all the time. Just as Darwin had observed how finches evolved in the Galápagos to adapt to different food sources, modern biologists saw bacteria evolving to escape antibiotic treatment and molecules evolving to take on new functions. So when I gave my talk, I was expecting provocative questions such as, "How do you think the *Tetrahymena* intron got into the gene in the first place?" I had considered such questions, I was prepared to speculate, and others in the Evolutionary Group would have chipped in their own ideas. But instead, I got questions such as, "Do you think your ribozyme could explain how life got started on the planet?" I hadn't thought much about the origins of life, so I felt totally unprepared to say anything useful. I ended my seminar quite mystified by the questions being tossed at me.

For one thing, I hadn't even known that there was a community of scientists pondering such primordial events. Nor did I understand why they were so animated by my research. One of the participants was Bill Schopf, a UCLA professor of paleobiology, who was finding microfossils of cell-like structures in very old rock. He would soon announce that a rock from Warrawoona, Australia, seemed to contain single-celled life-forms dating from 3.3 to 3.5 billion years ago. The big question was what had been inside those cells: DNA, RNA, or something else entirely? Unlike the exteriors of the fossilized cells, the molecules in the interior are too small to retain their shape when they're turned into rock. Otherwise, they would have given Schopf clues about their composition and structure.

The idea that fossils could provide evidence for ancient events was certainly not new to me. As a fourth-grade student at Dr. Howard School in Champaign, Illinois, I collected fossilized shells and snails—once-living creatures now entombed in limestone. My paperback *Guide to Rocks and Minerals* showed how fossils were sometimes encased in iron concretions, and when I found a similarly shaped rock, I stood it on its edge, smacked it with my Estwing rock hammer, and—much to my amazement—a perfect fossilized fern revealed itself. It had been waiting 300 million years for me to come along. So I was comfortable thinking about fossils from the Carboniferous Period, but the very origins of life on Earth had somehow eluded my attention—until that evening at UCLA.

After returning home from L.A., I began reading some old scientific papers. Then it hit me. For decades, scientists in the origins-of-life community had been pondering how the first self-reproducing system might have gotten started on Earth, almost 4 billion years ago. And when they encountered the chicken-or-egg problem, they had already theorized that RNA might offer a solution.

Clearly, RNA was an informational molecule: acting as a messenger for DNA, it carries the code that directs the order in which amino acids are laid down in a protein; and in RNA viruses, RNA is the repository of all the genomic information needed by the virus to carry out its infectious cycle. So there was no question that RNA could carry information, the instructions necessary to kick-start life. The question was how it could be replicated or reproduced in a protein-free primordial world. Without replication, a new generation of RNA can't be created from the previous one.

One scientist working on this problem was a British chemist named Leslie Orgel, who by that time was on the faculty of the Salk Institute in La Jolla, California. Since the 1960s, Leslie had been pining for DNA or RNA molecules that could reproduce themselves without a protein enzyme, resolving the chicken-or-egg problem. Yet in an oft-cited 1968 paper, he had said there was "no evidence" that such a molecule had ever existed, and he expressed doubt that a primitive form of RNA could have had the power to crack open the book of life.

But, he added, tantalizingly, "One cannot be quite sure."

In this paper, Leslie had been dancing around the idea of RNA catalysts. This was why the Evolutionary Group at UCLA—and, as I would soon discover, Leslie himself—was enormously excited by our discovery of ribozymes. *RNA indeed had information and function in the same molecule!* But it was even better than that. The reactions catalyzed by our self-splicing intron all involved the creation of new chemical bonds between RNA nucleotides.* That was exactly the sort of activity that a ribozyme replicase would need to possess to achieve RNA self-replication. Perhaps in the beginning there was only RNA, and proteins and DNA came later.

Despite my being so mystified that night at UCLA, I now found myself getting more and more interested in these origin-of-life questions. Throughout the 1980s, I would have many animated discussions

---

*The three reactions included the addition of a guanosine to the intron, the joining of the ribosomal RNA sequences that had been interrupted by the intron, and the tying of the cutout intron into a circle.

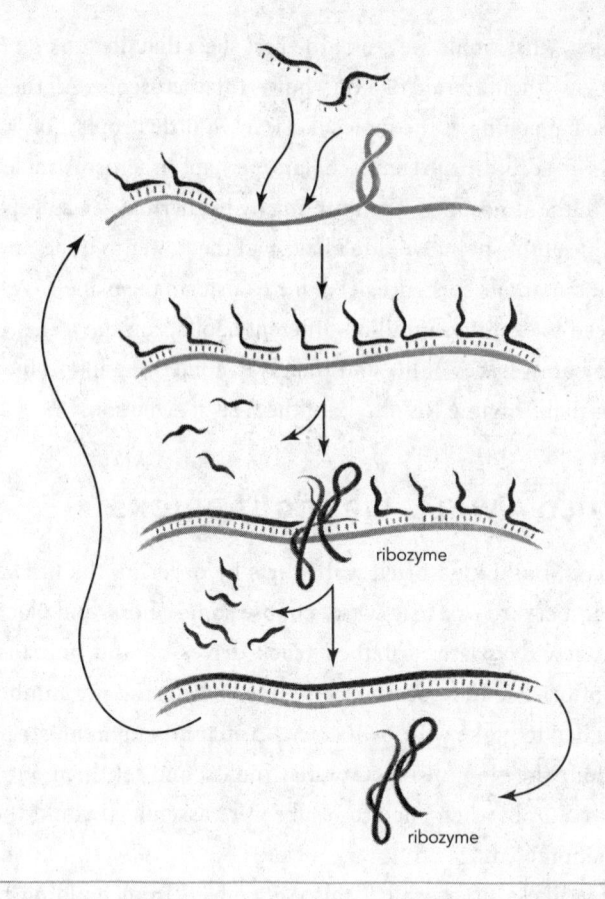

Hypothetical RNA self-replication in which small scraps of RNA bind by base-pairing to a preexisting RNA strand (light-shaded) and are then stitched together by a ribozyme (dark-shaded). The product is a double-stranded RNA. The energy provided by sunlight then "melts" this double-stranded RNA, separating it into its two strands. The dark strand folds within itself to form a new ribozyme, while the light-shaded strand provides a template for another round of replication.

with Leslie in his office overlooking the Pacific, considering the plausibility of ribozymes copying themselves.

But as exciting as the idea of such an "RNA world" seemed to chemists like Leslie and me, I must admit that such a place would not have been very exciting to visit. If we could be transported back in time to take a look, what would we see of the first steps that life took on Earth? Very little, without the aid of a very powerful microscope. All the action would be happening at the molecular level in little droplets of water in the rocks or, according to some scholars, perhaps in aerosol droplets suspended in the atmosphere or in fuming hydrothermal vents deep in the ocean. Life at this point wouldn't have had the power to build anything that would mark its existence, let alone transform the planet. We would likely need to examine a million different habitats on the planet to find even the barest traces of life and then wait something like 100 million years for that fledgling RNA to light the fuse of evolution.

TO BUILD A WALL, FIRST GET BRICKS

These days, building a brick wall starts by ordering the bricks. You visit a lumberyard or a brickyard, choose some colors, and place your order. A few days later a flatbed truck drives up and unloads your pallets of bricks, and you start building. But in the pre-lumberyard era, you had to make your own bricks. You'd mix up mud, straw, and water, pour the glop into rectangular molds, and set them out in the sun to dry. Only when enough bricks were assembled could the wall construction begin.

RNA, like a brick wall, is also assembled from building blocks, the four nucleotides A, G, C, and U, each with three phosphate groups attached to chemically activate them and make them prone to joining with each other. So anyone who's trying to argue that RNA was indeed the first life-form on the planet needs also to come up with a satisfactory explanation for how the building blocks of RNA came to be.

Today, most researchers studying RNA self-replication buy their "bricks" (i.e., nucleotides) in pure form from a chemical warehouse and receive them by express mail. In prebiotic times—that is, before the emergence of the first life-form—the bricks would have to form spontaneously from chemicals present in the environment. These chemicals would be akin to the mud, straw, and water needed for brick-building.

The carbon, hydrogen, and nitrogen found in each nucleotide would come from simple atmospheric gases such as hydrogen cyanide, which contains all three of these elements. Hydrogen cyanide is thought to have been abundant in the atmosphere of the young Earth, and although it is poisonous to humans and other creatures, none of us were around back then. The oxygen needed for nucleotides would come from water. The fifth element needed to form a nucleotide is phosphorus, which would need to come from phosphate-rich terrestrial rocks called apatites or from extraterrestrial meteors bombarding the earth.

How feasible is it that the nucleotides that compose RNA could have formed spontaneously using materials present on the primitive Earth some 4 billion years ago? The pioneering chemists Stanley Miller and Harold Urey showed in 1952 that many of the amino acids found in modern proteins were formed when simple gases were ignited with an electric spark, which they used as a substitute for lightning. Might the "bricks" that form RNA, the nucleotides, also be formed under plausibly prebiotic conditions? Indeed, work in the lab of the British chemist John Sutherland has shown that simple chemical compounds containing nitrogen, oxygen, carbon, hydrogen, and phosphorus—compounds plausibly present on the early Earth—can react to form nucleotides.

But a vexing problem emerged: the reaction conditions needed to make the U and C bricks were quite different from the conditions needed to make the A and G bricks, and the two sets of conditions were largely incompatible. If we're going to build a wall with four colors of brick, we need to collect them all in the same place. A breakthrough emerged from the lab of the German biochemist Thomas Carell in 2019. Starting with molecules that could have plausibly existed on prebiotic Earth, Carell and his colleagues found that cycling between wet and dry conditions allowed all four nucleotides to accumulate essentially in one pot. It's quite likely that the environment on the early Earth did indeed cycle through wet and dry conditions: day followed night then as it does now, and droplets of water and dissolved compounds that condensed on rocks in the cool of the night would have begun to evaporate under the sun's rays. This evaporation would first concentrate the compounds in the water, which encourages chemical reactions, before the droplets dried up entirely. This isn't to paint too gentle of a picture of our planet in those distant days. Earth was a violent place back then, full of lightning storms, comet bombardments, volcanic eruptions, and powerful UV radiation from the sun. This harsh environment contributed the energy required to drive chemical reactions.

Let's accept that Sutherland, Carell, and the other scientists working on prebiotic nucleotide synthesis are correct, and that it's plausible that nucleotides would have formed spontaneously on the ancient Earth. Then we'd have our bricks, and we'd be ready to build our wall—the "wall" being a string of nucleotides with the ability to reproduce itself. In the early 1980s in his Salk Institute laboratory, Leslie Orgel performed proof-of-principle experiments. He synthesized nucleotides and showed that if he stewed them up at very high concentrations and waited for days, they would spontaneously react with each other to form short strings of RNA. If he started out with the C nucleotide, the reaction products would include CC, CCC, CCCC, and CCCCC. When he then added G nucleotides, they would line up on the strings of C by C-G base-pairing and react to form short strings of G's.

Although it was exciting to see even these short scraps of RNA being formed without any enzyme to power their assembly, the reaction was extremely slow and inefficient. Ribozymes may have been the missing ingredient. What if one of Leslie's randomly produced scraps of RNA happened to be long enough and to have the right nucleotide sequence to be able to fold up into a catalyst that could copy itself? Then, instead of growing in fits and starts, the RNA could reproduce itself whole cloth, in the kind of replication that would have allowed RNA to be the miracle molecule that catalyzed life on Earth.

The odds of this happening spontaneously are probably a lot lower than your winning the Mega Millions lottery. But then again, the prebiotic RNA game would be played in a million locations all over Earth. If it took 100 million years to find a winner, no problem—life could wait.

BUILDING THE WALL

Is RNA-catalyzed RNA self-replication in fact realistic? Is it possible to reproduce it in the laboratory? That would at least demonstrate its feasibility, if not prove that RNA was actually the springboard for all living things.

In January 1986, two years after that evening seminar at UCLA, I was invited to the University of California, San Francisco, to give a talk. I was having lunch with the chair of the biochemistry department, Bruce Alberts, at a café on Parnassus Heights above Golden Gate Park. Bruce, who would later go on to become president of the National Academy of Sciences, would often invite scientists to come to San Francisco and share their recent research findings.

"Jack Szostak was here last week," said Bruce between bites of his pastrami sandwich. "He's really excited about origins-of-life research, and he's redirecting his entire research program to study your ribozyme."

I came close to aspirating an olive from my salade niçoise. Jack Szostak, a young Harvard professor, already had a huge reputation for deciphering the fundamentals of DNA recombination, or the exchange of DNA sequences from one chromosome to another. On the one hand, I was excited that a geneticist as highly regarded as Jack would work on our ribozyme. On the other, I felt a frisson of terror knowing how creative and productive he was. Would Jack do every experiment I wanted to do, but faster?

Jack thought that the origin of life was the greatest unanswered question in science. He believed then—and still does today—that if one can approximate the conditions of the prebiotic Earth in the lab and achieve RNA self-replication, then one can be reasonably sure how life got started on our planet.

Jack had a secret weapon: Jennifer Doudna. Before she came to my lab as a postdoc, Jennifer did her graduate work with Jack at Massachusetts General Hospital. The grand goal of her PhD dissertation project was to go beyond the type of reactions Leslie Orgel had achieved, which made random scraps of RNA, and instead replicate an RNA that served a useful purpose. She aimed to get a ribozyme to make a copy of itself in the test tube, thereby proving the feasibility of one of the key steps needed for RNA self-replication. When he got to work in the morning, Jack would head straight for Jennifer's bench—that's where the action was. And often she had a new breakthrough to share.

In nature, the RNA cutting-and-pasting reactions catalyzed by the *Tetrahymena* ribozyme all take place within a single chain of RNA. Starting in 1986, my research group showed that the intron part of the ribozyme could catalyze the cutting and pasting of separate RNA molecules—a first step in creating an RNA replicase. But our system was limited in the RNA sequences that it could handle, so it lacked the versatility that would be needed for RNA self-replication. In 1989, Jennifer and Jack announced a breakthrough: they had engineered the *Tetrahymena* ribozyme to copy longer separate RNA strands of diverse sequence.

To understand the origins-of-life importance of this achievement, we have to go back to the basics of how nucleic acids—both RNA and DNA—are copied in nature. That process is never direct. A single strand of nucleic acid doesn't simply duplicate itself, as in GGG → GGG. It first needs to be copied into a template, a complementary strand, and only then can the magic of complementary base-pairing be used to direct the synthesis of another copy of the original molecule. In other words, if you want to reproduce GGG, you first have to copy it into CCC, which can in turn be used to direct the formation of another GGG: GGG → CCC → GGG.

The process works a bit like casting a 3D object from a mold. Let's say you have a plaster garden gnome and want to make yourself a twin. First you have to make an inverse replica to use as a mold. All the structural details of your gnome—his long beard, pointy hat, and round belly—become concavities in your mold. Once you have your mold, you can pour in plaster and cast an identical replica of the original gnome. In the RNA self-replication scenario, a ribozyme with replicase activity would be the garden gnome, and the mold would be an RNA with its complementary sequence. This complement would have no catalytic activity, but it would be required as a template to make more ribozymes. Finally, copying the template strand back into a ribozyme would be akin to pouring plaster into the mold to make another gnome.

What Jennifer and Jack achieved in 1989 was reengineering the *Tetrahymena* ribozyme so that, when given a template RNA, it catalyzed the construction of a complementary strand. Starting with a mold, they could make another gnome. Their reengineered ribozyme could copy all kinds of RNA sequences, but the ones that would be relevant to RNA self-replication would be the ribozyme sequence and its complement: ribozyme → complementary sequence → ribozyme. So they had re-created one key step in the journey from RNA building blocks to RNA self-replication.*

But there was another gap in the RNA self-replication journey that Jennifer and Jack wanted to address: the longest strand they'd been able to make was 42 nucleotides—a world record at the time, but far short of the 400 nucleotides needed to make the *Tetrahymena* ribozyme itself. They could only make the head of the gnome, not the whole creature.

To overcome this size limitation, Jennifer took a two-pronged approach. First, she switched from *Tetrahymena* to a different ribozyme. Researchers at SUNY Albany had recently found a bacteriophage ribozyme, SunY, that had the same sort of self-splicing activity as the *Tetrahymena* ribozyme but was about half the size, so it would be easier to replicate. Half the battle was won. Jennifer then decided to divide and conquer: she cut the SunY ribozyme into three pieces that would find each other in the test tube and assemble by base-pairing. These fragments were now small enough that the SunY ribozyme was able to copy

---

*This is an *intermediate* step in RNA self-replication because it doesn't address the question of how an RNA as sophisticated as the *Tetrahymena* ribozyme would arise in the first place from random chemical reactions. It's difficult to conceive of lab experiments that could recapitulate those early steps, which might have taken 100 million years on the early Earth.

them. Thus, Jennifer and Jack had shown that it was feasible for fragments of RNA—the sort that might arise spontaneously by Leslie Orgel–type reactions—to assemble with each other to form a little machine capable of self-replication.

So, scientists have been able to reenact in test tubes many of the steps that would have been required for prebiotic RNA self-replication—making nucleotides, stitching them together into RNA molecules, and finding a ribozyme that could assemble a copy of itself on a separate RNA molecule. Although scientists have not yet succeeded in producing an entire RNA self-replication cycle—mixing RNA nucleotides in a test tube and coming back later to find an assembled RNA molecule busily making copies of itself—the proposal that life started with a primordial RNA world at least seems plausible.

PLEASE ENVELOP ME

Researchers were building a strong case that life on Earth could have originated in an RNA world, but they still had a major problem that needed to be resolved. A bunch of molecules in a drop of liquid is, of course, not an organism. An organism, even a primitive one, needs to be an entity distinct from its environment and distinct from other organisms. It needs to be encased in some sort of envelope.

Just as plastic wrap protects your tuna sandwich from falling apart and getting dirty, animal cells are surrounded by a membrane that protects them from at least some of the dangers lurking outside—toxins in the environment as well as bacteria, viruses, and other pathogens. Clearly this protection is incomplete, and some of these invaders get into our cells, but the vast majority are warded off. These *cell membranes* are made of *lipids*—fatty molecules that have the wonderful properties of self-assembling into two-layered sheets strong enough to withstand pressure, impermeable enough to protect the cell contents, and yet pliable enough to allow cell movement and cell division.

In a primordial RNA world, an ancient cell would similarly benefit from a membrane that would surround and protect it. For example, a membrane could exclude competing RNA molecules. Just as dogs get fleas, which benefit from their location but do the dog no good at all, replicating RNA molecules suffer from parasitic RNAs that come along for the ride, stealing nutrients from the self-replicating molecule without doing it any good. We see these popping up in test-tube evolution experiments in the lab, and they would inevitably arise in nature. A membrane would keep the self-replicating RNA inside and would keep parasitic RNAs on the outside from getting in.

Furthermore, envelopes facilitate evolution. If multiple self-replicating RNA molecules are mixed together in the same drop of water, then any mutation that makes one molecule work better as a replicase will benefit all of the molecules in the vicinity. Although such altruistic behavior may seem admirable, it inhibits evolution. For life-forms to improve over time, there needs to be "survival of the fittest," and only when life-forms are separate and distinct from one another can an entity benefit from the acquisition of a favorable mutation and win out over the others. This may sound Machiavellian, but at least when it comes to the evolution of the species, individual selfishness benefits the community in the long run.

Jack Szostak's research group has been examining how nucleic acids behave inside of membrane envelopes in what he calls "protocells"—artificial approximations of what a primitive cell might have looked like. It's easy to trap a nucleic acid such as RNA inside a protocell. You first mix the fatty acids that form the protocell together with the nucleic acid, then let the mixture dry out and rehydrate it or else subject it to freeze-thaw cycles, and nucleic acid is randomly encapsulated. Jack has seen that nucleic acids can form longer chains in a Leslie Orgel–type reaction within the protocells. His group has also shown that these protocells can grow and then divide, although not with the regularity with which modern cells undergo cell division. Protocells provide one more step toward achieving RNA self-replication in the laboratory, giving us a plausible scenario for how the inorganic could have morphed into the organic.

Scientists are getting closer to proving, at least in the lab, that life *could* have started in an RNA world. They've managed the impressive trick of getting RNA to build itself in a test tube, but they have a long way to go before they reach their ambitious goal. They still need to figure out how complete ribozyme self-replication could occur in a protocell, in the climatic conditions of the prebiotic Earth. And they need to watch as those protocells divide, and mutate, setting off the kind of evolutionary event that could have been the springboard for the origin of life.

But even if chemists succeed at all this in the laboratory, a fundamental issue will remain: the origin of life is less a scientific question than a historical question. Just because RNA *can* self-replicate wouldn't prove that it did, thereby kicking off the whole evolutionary process that led to life on Earth as we know it.* Will we ever know if it truly was RNA, or some cousin of present-day RNA, that inhabited Bill Schopf's fossilized cells? Is it even *possible* to know how life started on Earth almost 4 billion years ago?

When we look down into the kiva and try to peer into the darkness of the sipapu, we are in a way reaching out across the millennia, connecting with other humans who have pondered the origins of life. Because we live in an age of amazing science, we may think we can make out the shape down there in the darkness, and we have good reason to believe that shape looks a lot like RNA. But we are not quite certain. Science can *suggest*. Science can say "it is plausible." But science will probably never be able to *prove* whether life started with RNA.

The impossibility of such proof has always made me a little uneasy about origins-of-life research, which is sometimes spun in speculative

---

*A competing theory posits "proteins first," and indeed proteins may have some ability to direct the synthesis of new protein molecules. However, it's quite obscure how such a "protein world" would transition to nucleic acids as informational molecules, whereas the RNA world could form a primitive ribosome to synthesize proteins.

directions that verge on hype. I remember confessing my discomfort to Leslie Orgel during one of our discussions at the Salk Institute.

"As long as you are uncovering fundamental principles of nucleic acid chemistry," he said, "then the research is valuable, and origins of life is an interesting hook to hang it on." I always thought that was well said. Of course, Leslie said it in his wonderfully authoritative British accent, which may have influenced my willingness to accept his advice.

Ultimately, the origin of life may be the deepest question provoked by studying the nature of RNA. But as intriguing as it is to contemplate RNA's contribution to life's deep history, it's time to return to the ways that RNA is reshaping our present and future. RNA is already catalyzing a revolution in medicine—and, as we'll see, it has the potential to extend healthy lives beyond nature's current limits.

PART II

THE CURE

Chapter 7

IS THE FOUNTAIN OF YOUTH A DEATH TRAP?

My university office has a knickknack shelf full of souvenirs collected on conference trips and little gifts from former students. Sitting among these is a green plastic bottle of pills that someone thought I'd enjoy as a conversation piece. Its label boldly promises "Cell Rejuvenation through Telomerase Activation."

These pills are among the many newfangled supplements that seek to profit off the halo of "immortality" associated with an RNA-powered enzyme called *telomerase*. On Amazon, you can buy a "telomere-protecting" antiaging cream called "Youth Shots" for the seemingly reasonable price of $25.99. Meanwhile, "HealthyCell Telomerase Activator" capsules have garnered 400 five-star reviews, including one by a customer claiming that the product cured his mother's Alzheimer's disease. Another reviewer noted that it also "tastes great."

I find it decidedly weird to see how this enzyme—telomerase—has risen from an arcane scientific topic to a buzzword in just a few decades. In the 1980s, telomerase was of interest to a small group of us who were studying pond scum. Now it is marketed as a veritable Fountain of Youth, part of the multibillion-dollar antiaging industry. You might think that the pursuit of immortality is a pipe dream reserved for quixotic billion-aires. But on the cellular level, at least, immortality already exists. And telomerase is the secret sauce that makes it possible.

Built out of proteins and RNA, telomerase enables cells to keep dividing by adding protective genetic material to *telomeres*—the ends of chromosomes. Chromosomes are like little strings of DNA pearls nestled inside a cell's nucleus. In the absence of telomerase, the pearl at the end of the string is lost each time a cell divides; the whole string becomes slightly shorter. This process of attrition eventually leads cells to stop growing and enter a state called *senescence*, the cellular equivalent of old age. But telomerase forestalls this process. It adds pearls to the ends of the chromosome string, preventing senescence and rendering cells forever young.

Telomerase is produced by the rapidly growing cells in the human embryo, but this action gets turned off in most of our cells by the time we're born. The few key exceptions include stem cells—a true wonder of nature. Stem cells divide asymmetrically, meaning that, unlike most cells that produce two identical copies of themselves when dividing, stem cells produce offspring that differ from each other. One "daughter cell" of a stem cell will become a new stem cell, just like its parent, while the second will become a cell that the body needs to replenish, whether in our skin, our bloodstream, our hair, our digestive system, or other internal organs and tissues. The controlled proliferation of stem cells allows the human body to renew itself, and this vital process would not be possible without telomerase. Yet while telomerase is key to the proper functioning of stem cells, it's also a hallmark of most cancers. Whenever tumor cells stumble on a way to restart the production of telomerase, they escape the normal cell-aging process and achieve immortality—which often has lethal consequences for us.

So, is telomerase a miracle or a curse? Because this RNA-powered machine gives cells the ability to divide continually instead of senescing, it's only natural to wonder whether one could somehow harness its power to extend the vitality not of a single cell but of an entire organism. Could some drug based on telomerase actually keep our biological clocks ticking? To begin to answer that question, we have to turn back to my favorite single-celled furball, *Tetrahymena*.

ANOTHER LESSON FROM POND SCUM

Back in 1977, while still a postdoc at MIT, I drove from Cambridge to New Haven in my old stick-shift Volvo to spend a day visiting the lab of Yale professor Joe Gall. At that time I was just waking up to the possibilities that the microscopic critter *Tetrahymena* might provide to a researcher, and I wanted to visit Joe's lab because he had recently discovered an unusual set of *Tetrahymena* genes that existed as *minichromosomes*, each less than a thousandth the size of the smallest human chromosome. These individual DNA molecules harbored *Tetrahymena*'s ribosomal RNA genes. A few short years later, they would lead to RNA self-splicing and the discovery of the first catalytic RNA molecule—but none of that was even on the radar screen yet.

After a morning in Joe's lab staring through a microscope, watching *Tetrahymena* scoot around the limited environment of a glass microscope slide, it was time for lunch. Joe's group took me to the café on the top floor of Kline Biology Tower, the tallest building on the Yale campus, whose design clearly eschewed the principle that horizontal connectivity stimulates interaction and collaboration. But we were all far more interested in scientific discussion than architectural criticism. Joe's group was abuzz about the research of their Australian postdoctoral fellow, Liz Blackburn.

Liz had grown up in Hobart, on the island of Tasmania. Bitten by the science bug at a young age, she followed her interest in biology all the way to graduate school at Cambridge University. There, working with the two-time Nobel laureate Fred Sanger, she sequenced the DNA of a bacterial virus—a cutting-edge accomplishment at the time. This expertise proved to be the perfect background for taking on an adventurous new assignment as a postdoc at Yale, where she quickly made progress determining the DNA sequence of the ends of the *Tetrahymena* minichromosomes.

At the time, Liz wasn't thinking about breaking new ground in understanding cancer or the aging process. Nor was she thinking about cracking open a new chapter in RNA science; she just thought she would be telling another DNA story. None of us knew what sort of DNA might reside at the very ends of chromosomes, the linear DNA molecules that reside in the cell nucleus of any organism. Cell biologists had a longstanding interest in these chromosome ends, or *telomeres* (literally "end parts"), dating back to observations in fruit flies by Hermann Muller and in corn by Barbara McClintock. In 1938, both McClintock and Muller reported that if a chromosome breaks, as can happen spontaneously in nature or can be induced with X-ray radiation, the broken chromosome ends become unstable, either fusing with other broken ends or degrading. In contrast, the natural ends of chromosomes were somehow protected from these fates. Just as shoelaces have little plastic sheaths, called aglets, at their ends to keep them intact and prevent them from unraveling, chromosomes have telomeres. But for the next 40 years, no one had figured out what it was that allowed the chromosomal telomeres to work as aglets.

A thousand laboratories around the world were working on sequencing the middle parts of chromosomes—the parts that contained the genes—but the ends remained almost completely unexplored. So Liz and Joe focused their energies there. What did the DNA of the chromosome ends look like, and how were those ends protected? They decided to use the *Tetrahymena* minichromosome because its 10,000 copies per cell provided enough material to give them a fighting chance.

At the end of each *Tetrahymena* minichromosome, Liz discovered something very strange: a short six-letter sequence repeated many times. One strand had repeats of CCCCAA, and the other had repeats of the complementary sequence, TTGGGG:

```
TTGGGGTTGGGGTTGGGG . . .
AACCCCAACCCCAACCCC . . .
```

It was like reading a novel and encountering an otherwise sensible sentence that ended with *etc.etc.etc.etc.etc.etc.* One "etc." could make sense, but a long string of them would seem completely redundant. What could it mean?

Liz and Joe are now widely recognized for having determined the first DNA sequence of a telomere. But it's fascinating that, in their 1978 paper reporting their findings, the word *telomere* was never mentioned. They announced the very first telomeric DNA sequence, and they didn't say a word about it! Why so cautious? *Tetrahymena* minichromosomes were so unusual—so much smaller than human chromosomes and present in so many copies—that it would have seemed presumptuous of the authors to have claimed that the telomeres of large normal chromosomes would be similar. This happens often in science. If your work is way ahead of its time, it takes a while for anyone—including you—to appreciate its full significance.

BIG CLUES FROM TINY ORGANISMS

It would take another decisive experiment to convince Liz Blackburn that she had unlocked the key to telomeres. In 1978, she moved to the University of California, Berkeley, to start her own lab as an assistant professor. At a conference in New Hampshire in 1980, Liz struck up a conversation with Jack Szostak, then a new faculty member at the Dana Farber Cancer Institute in Boston. Jack was studying the chromosomes of baker's yeast. He had found that he could sneak artificial circles of DNA into the yeast cells, and they would be maintained there as minichromosomes, but that linear DNA molecules treated the same way couldn't survive. This seemed backward because yeast's natural chromosomes are linear DNA molecules, not circles.

Jack and Liz wondered whether the linear DNA molecules were unstable in yeast because they were lacking some special stabilizing feature at their ends. Perhaps these shoestrings needed aglets. The only known DNA aglets were the ones that Liz had found at the ends of the

Tetrahymena minichromosomes. Was it possible that they could serve this stabilizing function in yeast?

In 1982, Jack and Liz collaborated on what was certainly a long-shot experiment. They transplanted the *Tetrahymena* DNA termini—the TTGGGG repeats—onto the ends of a piece of yeast DNA. And their hunch proved right. The *Tetrahymena* DNA ends allowed the linear DNA to be maintained stably in yeast. This was particularly astounding given the vast evolutionary distance spanned by these organisms: *Tetrahymena* is about as far removed from yeast as it is from humans.

Janis Shampay, a grad student in Liz's lab, sequenced the ends of the linear DNA that was now stable in yeast. This was not guaranteed to be very interesting. She might have seen that the ends remained capped-off by the *Tetrahymena* TTGGGG repeats. But what she saw was remarkable. The DNA molecules no longer ended in *etc.etc.etc.etc.etc.* but rather *etc.etc.etc.etc.etc.vs.vs.vs.vs.* And the *vs.* sequence turned out to be the same one that yeast used to cap off the ends of its natural, full-size chromosomes. It was yeast's own telomeric sequence. So, even though *Tetrahymena* and yeast were totally distinct species, their telomeric sequences were similar enough that when yeast detected the imported telomere (*etc.*), it began adding its own brand of telomeric repeats (*vs.*) to the ends of the minichromosome.

Janis, Jack, and Liz could think of only one way to interpret these sequence results. The sequence repeats—*etc.* in *Tetrahymena* and *vs.* in yeast—must be acting as telomeres, conferring stability to the chromosome ends, keeping them from eroding. Furthermore, yeast appeared to have a telomere-extending enzyme that was recognizing the *Tetrahymena* repeats as a "seed" and adding its own telomeric sequences onto them. That meant there must also be a telomere-extending enzyme in *Tetrahymena* that was creating its own brand of repeats. It all fit together so nicely, but were they building a house of cards? The proof would come if they could actually find this hypothetical telomere-extending enzyme. And a new grad student in Liz's lab was up to the challenge.

TELOMERASE EXISTS—AND IT NEEDS RNA

Carol Greider had a tough time even getting into grad school. She was dyslexic and scored poorly on standardized exams. But the Department of Molecular Biology at UC Berkeley looked deeper than her test scores and was impressed by her college research, so they took a chance on her—very wisely, as it would turn out.

For her part, Carol was thrilled not only to be at Berkeley but also to join Liz's young lab, where she took on the ambitious task of purifying the still-hypothetical telomere-extending enzyme from *Tetrahymena*. If indeed it existed, the enzyme would be able to add TTGGGG repeats to DNA ends. This was a risky project for a beginning PhD student—to find something that had never been found before, which, by the way, might not even exist. Little did Carol imagine at the time that the payoff would be not only the three letters "PhD" behind her name but also a share in a Nobel Prize—and a window into the deep question of immortality.

Carol joined Liz's lab in May 1984. She immediately started growing *Tetrahymena* in 1-liter glass bottles, breaking the cells open and isolating their nuclei. After all, telomere elongation took place in the cell nucleus, so this was the likely place to look for the enzyme that catalyzed the elongation. She then froze and thawed the nuclei, which caused them to burst open and release their contents. The goal was to isolate the secret sauce that extended the ends of chromosomes after cell division.

Carol and Liz figured that a whole chromosome would not be needed to trigger the activity of the enzyme, but merely the end of the telomere, where the action happened. So Carol synthesized short DNA strands composed of repeats of TTGGGG (the *Tetrahymena* telomeric sequence), hoping that would be enough for the enzyme to recognize the telomere and extend it with additional repeats. She then incubated these artificial telomeres in test tubes with the broken *Tetrahymena* nuclei. By Christmas 1984, Carol was thrilled to see that the DNA was being extended in a six-nucleotide repeating pattern—TTGGGG repeats, one after another. She had found direct evidence for the enzyme that would later be dubbed *telomerase*.

As is often the case in science, answering one big question immediately leads to the next. How could a protein enzyme possibly know how to make a specific DNA sequence as long as six nucleotides? No such enzyme had ever been found. DNA and RNA polymerases are capable of synthesizing long strings of nucleotides, but they don't do it by themselves—they use DNA as a template. *Reverse transcriptases*, such as those found in retroviruses, use RNA as a template to make DNA. So Carol and Liz wondered whether there might be an RNA that acted as a template for the addition of TTGGGG sequences. After all, the power of complementary base-pairing would make it easy for RNA to "remember" TTGGGG; it would simply use A's to specify the T's and use C's to specify the G's. To test this idea, Carol set about pretreating the *Tetrahymena* telomerase preparation with ribonuclease (RNase), the RNA-degrading enzyme, to see whether it would make any difference.

I happened to be in Berkeley, giving a departmental seminar, on the day in January 1986 when Carol performed the experiment. During a meeting I had with her and Liz that morning, Carol told me about her idea to test for an RNA component in telomerase. Now that I was an "RNA guy," I was excited by the possibility that RNA might be performing yet another magic trick. Throughout the day, as various faculty walked me around the department to my scheduled appointments, I would stick my head into Carol's lab and ask her how her experiment was going. She and I were having a bit of fun, because it takes at least a day to do such an experiment, so it was unlikely she'd have something new to report every half hour.

After I returned to Boulder, I learned Carol had indeed found that the telomerase activity was destroyed by RNase treatment. Because almost all enzymes are proteins and have no RNA, they are unfazed by RNase treatment. But telomerase activity appeared to require RNA. Thus, telomerase became the newest addition to the short list of exceptions to the "all enzymes are proteins" rule. There was our *Tetrahymena* ribozyme,

IS THE FOUNTAIN OF YOUTH A DEATH TRAP?

the related self-splicing RNAs in other species, ribonuclease P, and the ribosome protein-synthesis machine. And now telomerase.

A few years later, in 1989, Carol would identify and sequence the RNA component of *Tetrahymena* telomerase. By then, she had graduated from Berkeley with her PhD and moved on to Cold Spring Harbor Laboratory, Jim Watson's famous beacon of biological research on the shores of Long Island Sound. Lo and behold, the RNA contained a stretch of AACCCC sequence that could code for the TTGGGG's at *Tetrahymena* telomeres—validating her and Liz's hunch that an RNA template directs what DNA sequence is added to chromosome ends.

The corresponding human telomerase RNA was soon also iden-

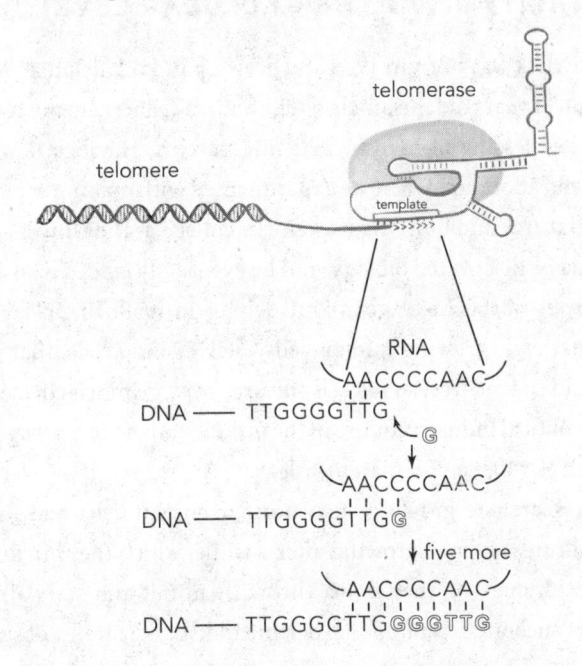

Telomerase uses a short portion of its RNA strand as a template, directing the sequence that's added to the end of the telomeric DNA. Nucleotides are added one at a time with the help of a protein (shaded oval). Shown is the *Tetrahymena* telomeric sequence with its repeats of TTGGGG. Once a complete telomeric repeat is formed, the DNA can slip back along the RNA, making room for the next repeat addition (not shown here).

tified and observed to act as a template for repeats of a similar sequence—TTAGGG—that composed human telomeres. So now RNA had been found at the heart of yet another critical life process—building out chromosome ends to secure the integrity of the genome.

All the research aimed at understanding telomeres and telomerase had been driven by curiosity, by the urge to know how chromosomes worked at a fundamental level. Initially, no medical applications were on the horizon. But this was about to change, as other evidence mounted that telomerase is central to both cancer and aging.

IMMORTALITY... AT THE CELLULAR LEVEL

Leonard Hayflick was born in 1928 and grew up in Philadelphia. When he was about 10 years old, his uncle bought him a Gilbert chemistry set, and, with the blessing of his very trusting parents, Hayflick built his own basement laboratory where he experimented with making explosive chemical mixtures and building rockets. In college at the University of Pennsylvania he discovered biology, and he eventually landed a position at the nonprofit Wistar Institute in Philadelphia in 1958. There Hayflick became a master at growing human cells, such as lung cells, that were virus-free and cancer-free, so his cell cultures became much coveted by the pharmaceutical industry for use in the production of vaccines against diseases such as rubella (German measles).

Other researchers growing such normal human cells had found that their cultures stopped growing after a while, which they attributed to sloppy technique—so they'd just throw them out and start afresh. Hayflick was such an exceptional experimentalist and careful observer that when his cultures stopped growing, he knew they were telling him something: normal human cells can divide a limited number of times, typically 50 to 60, before they enter the state of senescence. Senescent cells are not dead—they change shape, they switch their metabolism, and they keep living but simply don't divide. We now say that such cells have reached the "Hayflick limit."

Hayflick always believed that the limited proliferative life span of normal human cells made perfect sense. Just as it's critical for cells in the skin, liver, bones, and brain to keep dividing in an embryo and a child, it's critical for them to stop dividing in a full-grown person. This is especially true because the alternative—endless division—is the main hallmark of cancer.

But who's counting how many cell divisions a cell has undergone? There must be some sort of clock. The discovery of telomerase gave rise to the idea that telomere length might set the clock for the Hayflick limit. If telomerase was shut off in most human body cells, then the incomplete replication of telomeres would cause them to shrink, which could trigger senescence. In contrast, in continually growing organisms such as *Tetrahymena* and yeast—as well as in cancer cells—telomerase would always be "on," telomeres would maintain their length, and the Hayflick limit would never be reached. In 1990, cell biologist Cal Harley, who then had a lab at McMaster University in Canada, recruited Carol Greider to test the shrinking-telomere hypothesis. In a highly influential study, they found that as a type of human skin cells aged, their telomeres grew steadily shorter by about 50 base pairs per cell division. This correlation was intriguing, but Cal and Carol correctly concluded, "It is not known whether this loss of DNA has a causal role in senescence"—that is, whether it is actually responsible for the cessation of cell division.

Was telomerase indeed "the immortality enzyme" promoting long life? And when scientists found that increased telomerase activity is a hallmark of all kinds of cancer, did that mean telomerase would be a great target for cancer therapeutics? These proposed connections between telomerase, aging, and cancer brought biotech companies and Big Pharma into the hunt for the telomerase protein, because as critical as telomerase RNA is, it could only act in concert with its protein partner. Unlocking the secrets of telomerase required purifying the whole machine—RNA plus protein—which was extremely challenging because telomerase is rare even in the cancer cells where it makes its most terrible impact. Just a pinch of this stuff is sufficient to keep cells dividing over

chromosome

telomere

without telomerase, telomere shrinks with age

in cancer, telomerase is reactivated, telomere grows, tumor cells proliferate

senescence

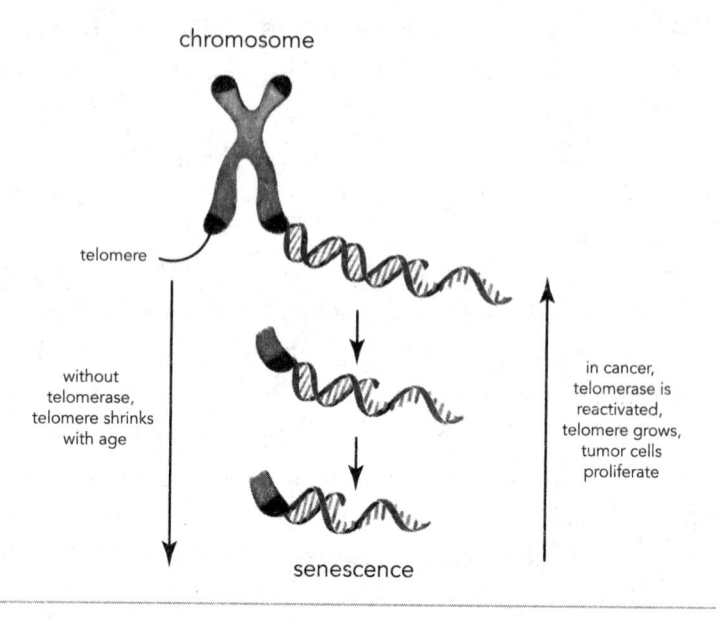

The telomere hypothesis for aging. Human cells require telomerase to maintain the length of their telomeres. Most somatic cells do not have telomerase, so their telomeres shrink as the cells divide. When telomeres become critically short, the cells quit dividing and enter senescence. Reactivation of telomerase is one of the steps required for cancer. Cancer cells are immortal and divide forever.

and over. Circumventing the challenge of purifying telomerase would require a Swiss postdoc named Joachim Lingner and another little pond critter distantly related to *Tetrahymena*.

RNA IS NOT ENOUGH

Basel, Switzerland, is a storybook city on the Rhine. It's the city where Switzerland meets Germany and France. It's a city of fabulous art museums—concrete walls displaying enormous Rothkos with stunning colors. Its five bridges span the Rhine, and between them are the city's four ferries, Wilde Maa, Leu, Vogel Gryff, and Ueli, which allow you to cross the river without motorized assistance. The ferries are ingenious, using the natural force of the river current to cross in one direction, and then with a flip of the rudder the same current is used to return. Equally ingenious is the science carried out in Basel—at the University of Basel, the Friedrich Miescher Institute, and two of the world's powerhouse pharma companies, Roche and Novartis.

In 1992, I traveled to the Biozentrum of the University of Basel to give a research seminar. During my visit I met a student, Joachim Lingner, who was finishing up his PhD studies under the direction of one of Switzerland's premier RNA scientists. Joachim asked if he could come to Boulder to purify telomerase. The enzyme would presumably contain an RNA subunit, as shown by Carol and Liz. And it would presumably contain one or more protein components to drive its DNA-extending activity. In 1993, I welcomed Joachim to Boulder, and I convinced him that we might be able to succeed where all the companies had failed by working with an organism that had an astounding knack for amplifying all things telomeric.

My organism of choice was *Oxytricha nova*, a critter that hung out with *Tetrahymena* in pond scum throughout the world. I had learned about *Oxytricha* from my colleague David Prescott, who had isolated multiple single-celled creatures from Varsity Pond on the Boulder campus. David had found something incredible: *Oxytricha* has 100 million truly tiny chromosomes, each harboring just a single gene. Because each chromosome has two ends, each cell contains 200 million telomeres. In contrast, humans have 23 pairs of chromosomes, and therefore 46 chromosomes or 92 telomeres in a typical body cell. Assuming that the amount of telomerase would scale with the number of telomeres, *Oxytricha* could give us more than a million-fold advantage over the teams of company scientists trying to purify telomerase from human cancer cells.

Like many "great ideas" promoted by research directors, my proposal had a few flaws. As Joachim soon found, it was difficult to grow many of these *Oxytricha*. We grew them in open lasagna baking dishes, which we bought from the local King Soopers grocery store, and they crawled along the bottom hunting for bacteria and algae. It was tedious to harvest the protozoa away from their food organisms, so Joachim decided to switch to a different species, a cousin of *Oxytricha* called *Euplotes aediculatus*, which was so huge by microbial standards (almost visible by the naked eye) that it stuck to cheesecloth while the bacteria and algae washed through. Culturing these creatures was still very time consuming, so we hired University of Colorado undergraduates to farm them. The undergrads would grow algae to feed the *Euplotes*, monitor them under the microscope to make sure they were doing well, and then move them to clean lasagna dishes as their population grew.

But how to purify the telomerase from these beasts? Joachim decided to identify its RNA subunit and use it as a "handle" to purify the intact enzyme. Working with an undergraduate student, he succeeded in isolating and sequencing the gene for the telomerase RNA subunit from *Euplotes*. The RNA of *Euplotes* telomerase was similar but not identical to that of *Tetrahymena*. That was just what we'd predicted—because these two RNAs were performing the same biological function but in different species, they had adapted and changed over the grand sweep of evolution.

Joachim's idea was then to go fishing for telomerase in broken-open *Euplotes* cells by using a short scrap of DNA complementary to the template region of the RNA subunit as a fishhook. The DNA fishhook should bind to the telomerase RNA template by complementary base-pair formation, and he could then pull the RNA out of the complex cellular mixture with its coveted proteins still attached. This worked beautifully. A year in the cold room—where biochemists work when they want to prevent damage to sensitive enzymes, just as we store our food under refrigeration to keep it fresh—and Joachim had achieved the first biochemical purification of telomerase from any organism.

Unfortunately, we had very little of the purified *Euplotes* telomerase, only about 10 micrograms. A microgram isn't very much—it's a millionth of a gram, and even a gram is only the mass of a raisin. We would have only one chance to get some protein sequences from this precious material, or it'd be back to the cold room for months. So, we needed a world-class collaborator. In the spring of 1996, we contacted Matthias

Purification of the telomerase protein took advantage of its RNA partner. The "bait" was a nucleic acid with a sequence that base-paired with the telomerase RNA template, thereby allowing the RNA and associated protein (shaded oval) to be captured and other cellular constituents left behind.

Mann, then at the European Molecular Biology Lab in Heidelberg, who had just invented a new protein-sequencing method and was keen to test it on an unknown protein. We sent him our irreplaceable telomerase and, in a short time, he sent us 14 snippets of amino acid sequence of the *Euplotes* telomerase protein—more than enough information for Joachim to isolate the corresponding gene.

One of the biotech companies infatuated with the connection between aging and telomere length was Geron Corporation ("Geron" as in "gerontology"), located in Menlo Park, California. They had been working nonstop to purify human telomerase from cancer cells but had found it very difficult. So, they hosted a conference, the Geron Telomerase and Cancer Symposium, at the Hapuna Beach Hotel on the Kona Coast of the Big Island of Hawaii, which took place over four days in

August 1996. Perhaps they hoped that the conferees, loosened up by an enchanting tropical location and a few mai tais, would spill some key information about the long-hidden protein that was partnering with RNA to drive telomerase.

At the symposium, I struck up a conversation over coffee with an old friend, Vicki Lundblad, then a professor at Baylor College of Medicine in Houston. Vicki had been one of my students when I was a graduate teaching assistant running general chemistry lab sections at UC Berkeley. Later, she had become a grad student of Jack Szostak, then a postdoc with Liz Blackburn. Vicki wanted to understand the fundamentals of how chromosome ends were maintained in yeast. Intriguingly, she had just discovered two new yeast genes that, when inactivated, caused the yeast to have *ever shorter telomeres*. She named the genes *Est* accordingly.

Yeast are single-celled organisms. Normally, they proliferate endlessly, and their telomerase is always active. In this sense, they are like human stem cells or cancer cells—dividing over and over. One plausible explanation for Vicki's new Est genes was that they encoded critical pieces of telomerase. Knock those genes out and the yeast telomeres shrink with every cell division, leading to cellular aging, or senescence. The DNA sequences of Vicki's Est genes matched nothing that had ever been seen before, so she didn't know where to go next. I told her about our *Euplotes* results, and we wondered if we might be hunting down the same target. As we talked about trading gene sequences, Titia de Lange, a famous telomere scientist from Rockefeller University, was pouring her coffee very slowly nearby. She had encouraged us to meet and was curious to hear the outcome.

While Vicki and I were in Hawaii, Joachim made a startling discovery back in Boulder. Staring at the sequence of the new *Euplotes* telomerase protein, he had the strange feeling of déjà vu. He had seen this sequence before, or at least something very similar to it, in the famous reverse transcriptase enzyme from human immunodeficiency virus. Why would our telomerase protein resemble a key protein in a virus such as HIV? The more Joachim thought about it, the more it made sense. Telomerase—like HIV—must use an RNA template to synthesize its DNA, and a reverse transcriptase protein could power that process.

When I got back from Hawaii, I connected Joachim with Vicki's student at Baylor, Tim Hughes. The goal was to compare the *Euplotes* and yeast gene sequences and see if there was some basis for moving ahead with a collaboration. And indeed there was. Yeast Est2 protein was a clear match to the larger of our *Euplotes* proteins, especially around the putative reverse transcriptase sequences. But unlike *Euplotes*, for which no molecular genetic tools were available, yeast would allow us to substitute multiple versions of a gene and see which ones worked and which failed to work.

A collaboration of furious intensity ensued. We constructed yeast Est2 genes with mutations confined to the reverse transcriptase sequences that Joachim had identified. I remember walking into the lab one afternoon and seeing one of my staff scientists holding a FedEx envelope as Joachim dropped in the tubes containing the DNA. He ran the package downstairs to intercept the last FedEx pickup. They were headed to Houston for analysis of their telomeres.

A few months later, it was all settled. Mutating even a single amino acid in the part of the Est2 protein that we hypothesized was powering telomere extension caused the yeast telomeres to shrink and the yeast to undergo senescence. They were aging, uncontrollably, right before our eyes. Thus, Vicki's yeast Est2 protein and, by extension, our *Euplotes* protein were critical to telomere elongation in living cells.

As thrilling as it was to have found some of the secret ingredients that prevented the aging process, our discoveries were, at least for the moment, confined to yeast and pond scum. Would these insights carry over to humans? Testing the relationship between telomerase, aging, and cancer would require the human telomerase protein, the critical partner to the RNA part that Carol and Liz had found.

These were the early days of the Human Genome Project, when new sequences of DNA were published pell-mell every day. Shortly before our *Science* paper was published, a piece of a human DNA sequence that was a close match to that of the *Euplotes* and yeast telomerase proteins appeared, unidentified, on our lab computer screen. It would be the key to finding the human telomerase protein. But once our paper appeared, others were sure to make this connection as well. We had a head start of only a few weeks over the rest of the world to isolate the human gene. The race was on!

One group hunting for the human telomerase gene was headed by one of the world's most renowned cancer biologists, Bob Weinberg of the Whitehead Institute at MIT. As fate would have it, I was on the Board of Advisory Scientists of the Whitehead. At the institute's annual retreat in the White Mountains of New Hampshire, I spoke to Weinberg postdocs Chris Counter and Matt Meyerson at their poster and learned that they were hot on the trail of human telomerase. I said something along the lines of, "You may need a new project, because we already have it." Not a smart thing to say to two superbly talented, highly ambitious postdocs, who immediately redoubled their efforts.

In the end, my lab won the race but not by much. Our paper describing the gene for human TERT (standing for *telomerase reverse transcriptase*) was published in *Science* on August 15, 1997. Bob Weinberg's group published a fine story on the human TERT gene in *Cell* just a week later. And indeed, mixing the RNA subunit with the TERT protein gave active telomerase, both in the laboratory and in living cells.

DO YOU PREFER AGING OR IMMORTALITY?

With both the RNA and the TERT protein components of human telomerase in hand, it was finally possible to test the idea that telomerase set the clock for the Hayflick limit. The first scientists to get this long-awaited answer were Prof. Woody Wright and his colleagues at the UT Southwestern Medical Center in Dallas in collaboration with scientists from Geron. They put the gene for TERT into normal human retinal cells, which were known to already contain the telomerase RNA but were missing TERT, and the cells now proliferated with no end in sight.

In contrast, retinal cells without TERT stopped dividing and showed the hallmarks of senescence after 50 to 60 population doublings. This provided compelling evidence that, in fact, shrinking telomeres are the yardstick that determines the Hayflick limit and that active telomerase prevents senescence. This trick is now used in biomedical research and industry to keep human cells from aging when grown in incubators. If you want your cultured human cells to keep proliferating without end, simply add the TERT gene.

The fact that telomerase can immortalize human cells, keeping them dividing continually in the laboratory without undergoing senescence, is a scientific fact. But it has been extrapolated, most unfortunately, to suggest that an increase in the level of telomerase could extend human life span. It's an overly simplistic idea: if our cells don't die, then we won't die either. This brings us back to my knickknack shelf with its "Life Extension" telomere creams and the "telomerase-activator" pills that are "clinically proven to lengthen telomeres." Because the ingredients in these pills and creams are natural plant products, they can be sold in the category of dietary supplements without being subject to the placebo-controlled clinical trials that are required for FDA approval of pharmaceuticals. They are *not* in fact "clinically proven."

But let's imagine for a moment that these pills and creams worked as advertised. What if they did keep our telomeres from shrinking and our cells from senescing? Would this indeed be a good thing? It's extremely difficult to imagine what would happen if all our cells divided continually. But, if one were to venture it, one outcome might be really large people, who kept getting larger without end. Or, given the relationship between continual cell division and cancer, perhaps these hypothetical telomerase-active people would succumb to a giant tumor.

Therefore, if it's to be beneficial, manipulation of telomerase activity and telomere length will need to be done more precisely. There are two situations in which this might have a lifesaving impact, if we can someday figure out how to turn this research into practical therapeutics.

The first situation concerns our stem cells, which have the job of replenishing worn-out cells in our bodies and therefore need to keep dividing throughout our lifetimes. There are rare individuals who are born with exceedingly short telomeres, shorter than 99 percent of those of other people their age. As a result, it doesn't take much telomere shortening for their stem cells to enter senescence, such that they can no longer maintain critical tissues. An inherited disease called dyskeratosis congenita arises exactly from this problem. The patients present with abnormal skin pigmentation, distorted fingernails and toenails, oral lesions, and tooth problems, and many later die from anemia. Further analysis reveals that they have a mutation in a gene for one or another of the components of telomerase, causing cells that need to keep on dividing to instead senesce. Likewise, many cases of a common blood condition, aplastic anemia, and a debilitating lung disorder, pulmonary fibrosis, are caused by too little telomerase and the shortening of telomeres that results. All these individuals would greatly benefit from a safe way to lengthen the telomeres of their stem cells. If we could develop an authentic telomerase-stimulating drug, the next challenge would be to target it mostly to stem cells.

The second situation is the flip side of the first. Most cancer cells begin as normal cells that have just a few fateful mutations that cause them to start dividing rapidly. In 90 percent of human cancers, telomerase is also reactivated and renders the cells effectively immortal. To give just one example of how enduring these tumor cells are, HeLa cells—part of the first immortal cell line obtained from the tumor of the famous cancer patient Henrietta Lacks in Baltimore in 1951—have active telomerase and are still alive in thousands of labs all over the world 70 years later. If all HeLa cells ever grown were placed end to end, it's estimated that they'd stretch for 350 million feet, enough to wrap around Earth three times.

Given the terrifying power with which tumor cells are endowed by telomerase, the hope would be to find a way not to enhance but to inhibit the telomerase in tumors or to prevent it from being turned on in the first place. But to do that, scientists would need to solve another mystery: how telomerase gets reactivated in tumors to start with.

A VERY SMALL CHANGE MAKES A VERY BIG DIFFERENCE

By the early 2000s, scientists all over the world had sequenced the TERT gene in tumors, but they could find no mutations that could explain how TERT was being turned on. That is, until Franklin Huang came along.

Franklin had grown up in Oklahoma, the son of Taiwanese immigrants. He obtained his MD and PhD degrees at Harvard Medical School, and in 2012 he began working as a medical fellow in Levi Garraway's lab at the Dana Farber Cancer Institute in Boston. Levi was a leader in the use of a powerful new technology from the company Illumina to sequence the DNA of tumors. Members of his lab were looking for mutations that might be driving cancer—and might also be targets for pharmaceutical intervention.

The lab had a vast collection of melanoma genome sequences, which had already yielded useful information about this skin cancer. Yet Franklin took a fresh look at the data. He quickly realized that something intriguing was going on in the gene called TERT. In 17 out of 19 melanoma samples, there was a single base-pair mutation in the same position. It was not in the coding region of the gene, where everyone else had been looking, but rather in the part of the gene called the "promoter"—because this part promotes transcription of the DNA into mRNA. It looked like the alteration would create a binding site for a protein called a "transcription factor," whose binding could plausibly drive transcription of the gene. Was this the genetic error that turned telomerase back on, giving these cancers room to run?

Franklin's colleagues in the lab were skeptical. They thought it was practically impossible for one cancer-causing mutation to appear with such high frequency. Maybe it was a mistake? Perhaps the high-tech

DNA sequencer had a problem with this particular sequence and misread it—a lot?

So Franklin sequenced the DNA samples by an old-school method—one that didn't involve the high-tech sequencer and therefore wouldn't succumb to its pitfalls, if indeed there were any. It took him all night to get it done, but the next day he had his answer. Most of the melanoma DNA sequences indeed had the mutation in the same position of the TERT gene that he had seen earlier. What's more, when he sequenced DNA taken from the same patients' blood, which wasn't cancerous, he discovered that none of the TERT gene sequences in that DNA had the mutation. In other words, the mutation was specific to the cancer, so it could not be a DNA sequencing error—it was real.

Franklin and Levi went on to show that the single-base mutation did in fact push the transcription of the TERT gene. Later, he and others found that many other cancers had also stumbled into this lucky-for-them trick of activating TERT, and thereby telomerase, through exactly the same mutation. Quite remarkably, this mutation arises independently, all around the world, hundreds of thousands of times per year. Presumably a host of other mutations arise with similar frequency, but they do not drive tumor progression, so they get diluted over time.

The discovery of the TERT promoter mutations has important diagnostic applications. For many types of cancers, the presence of a TERT promoter mutation signals a more aggressive disease—one that needs aggressive treatment if the patient is to survive. Searching for this mutation can therefore help physicians tailor treatment plans, perhaps recommending more conservative treatments in cancers without this mutation to avoid some of the debilitating side effects of chemotherapy.

While these diagnostic applications are already helping patients, the quest to turn this research into an effective therapeutic continues. One challenge is to inhibit telomerase in tumors but not in stem cells, which also rely on telomerase. As with all our RNA stories, knowing the biology—understanding the mechanisms—is essential for medical intervention, but it doesn't guarantee success. It's often a long and difficult path between making a scientific discovery and turning it into a cure, and for telomerase, the quest is still ongoing.

In the case of RNA interference, the next stop on our RNA journey, the turnaround between fundamental discovery and therapeutics happened a lot faster. Part of the reason why fundamental research is so important and exciting is that when we make a new discovery about the nature of RNA, we can never predict what medical applications might be waiting in the wings.

Chapter 8

AS THE WORM TURNS

SiQun Xu was wielding a worm-pick, deftly transferring minuscule roundworms from their Petri dish home to a pad of gooey agar laid down on a glass microscope slide. If you had been standing over his shoulder in June 1997, watching as he worked, you'd have been skeptical that anything was being plucked up and transported—the worms are transparent, thinner than a human eyelash, and only a millimeter long, so it takes a sharp eye even to see them. Once he had 10 worms lined up on the slide, SiQun peered through the eyepieces of the microscope, inserted an ultrathin glass needle through the first worm's skin into its gonad, and injected a tiny volume of dissolved RNA. Moving down the row, he injected each worm in turn. The process was much more difficult than threading a needle, but, years before becoming a research scientist in Andy Fire's lab at the Carnegie Institution in Baltimore, SiQun had worked as an acupuncturist in his native China; it was fortuitous training for the day's mission.

With these RNA injections, SiQun and Andy were hoping to solve a conundrum that was befuddling worm biologists. Antisense RNA—which we encountered as a therapeutic for spinal muscular atrophy—had been a popular tool for manipulation of gene expression since 1984.

The idea was to short-circuit the production of a protein by introducing an RNA that was complementary ("antisense") to its mRNA. The antisense strand would base-pair with the mRNA of interest, covering up the codons and preventing protein synthesis. The ability to precisely switch off genes could be a powerful tool to understand the function of different genes and perhaps even to deactivate harmful or mutated genes.

But when antisense RNA was applied to the roundworm *Caenorhabditis elegans*, a popular experimental organism, it did not have the expected result. A number of worm biologists had been exploring antisense RNA, and several of them—including Andy Fire—noticed something strange. As an experimental control, worm researchers injected sense RNA instead of antisense RNA—that is, a carbon copy of a portion of the target mRNA sequence. It should have had no hope of base-pairing with the mRNA because C does not pair with C, A does not pair with A, and so on. They fully expected that injection of sense RNA would have no effect. But most surprisingly, they found that sense RNA also interrupted gene expression. That sense *and* antisense generated the same result made no sense at all.

Andy thought up a possible solution to this riddle. He had been trained in Phil Sharp's lab at MIT, so he knew RNA. He also knew that it was challenging to make pure sense or pure antisense RNA, because the enzymes used in the lab to transcribe DNA into RNA sometimes made a mistake: in the process of producing the target RNA, they also generated some of the complementary strand. Was it possible that everyone's sense and antisense RNA preparations were active because they contained some double-stranded RNA?

Andy admitted that this was a "somewhat far-fetched hypothesis." After all, a double-stranded RNA was already paired up with itself—shouldn't that mean the strands would be unable to pair with anything else and therefore be unable to interfere with mRNA function? But worms were cheap, and SiQun and Andy were expert worm-injectors, so they decided to give it a shot. They would purify their sense and antisense RNAs very carefully to avoid any cross-contamination and then inject some worms with sense RNA, some with antisense RNA, and some with an equal mix of sense and antisense to form double-stranded RNA.

They decided to target a gene whose loss of activity would be readily apparent. The gene named *unc*, short for *uncoordinated*, was required for the worm's nervous system to develop properly. Mutation or inactivation of the unc gene causes worms to twitch uncontrollably. SiQun and Andy would look at the progeny of the injected worms, whose brains would either have developed normally or not. Conveniently, there was no need to coax the worms to mate, because they are hermaphrodites, making both sperm and eggs and internally self-fertilizing.

A day after their RNA injections, the worms had laid eggs, and the eggs hatched. As SiQun and Andy took turns looking through the microscope, they saw something thrilling and unexpected. First off, *only* the double-stranded RNA caused all the hatchlings of every injected worm to twitch like crazy. Neither the antisense RNA strand nor the sense RNA strand by itself did much;* only the double-stranded combination was active, suggesting that earlier findings that sense or antisense alone disrupted gene expression in worms may indeed have been the result of contamination by the opposite strand. Not only did they discover that double-stranded RNA could somehow knock down gene expression, at least in worms; they also discovered that it was impressively precise. It was only the targeted unc gene that seemed to be affected by the double-stranded RNA treatment.

It would take scientists a few years to explain these mysteries. But right then, as Andy Fire stared at the writhing worms, he was also staring at a Nobel Prize. His work, along with that of his collaborator, Craig Mello of the University of Massachusetts Cancer Center in Worcester, would launch an entirely new subfield of molecular biology called *RNA interference* (RNAi).

---

*Although the antisense RNA did not block gene expression in these particular experiments, it is active in many systems—for example, as developed by Adrian Krainer and Ionis to treat spinal muscular atrophy.

Scientists would soon reveal that RNAi was a key regulatory process in nature, allowing organisms to reduce the activity of groups of mRNAs after they are transcribed. This system, active in animals from worms to humans, had been flying under the radar until its signal was picked up by the double-stranded RNA experiments of Andy Fire and Craig Mello. RNAi offered another astounding example of how RNA was central to life. Furthermore, because RNA carries the message in every disease, just as it does in every healthy life process, the ability to knock down specific mRNAs had pharmaceutical potential—so RNAi would soon be redirected for medical use. That story shows both the great promise and some of the challenges that accompany the therapeutic use of RNA. And it all began with the lowly worm.

THE SOUNDS OF SILENCE

Why the worm? To most people, it would seem like an unlikely choice of experimental organism, perhaps even a jest. Not to Sydney Brenner. In the 1960s, after he had helped crack the secret of mRNA, Sydney turned his attention to one of the grandest remaining challenges in biology—understanding the nervous system. The nervous system consists of the brain, the spinal cord, and peripheral neurons that emanate from the spinal cord, including the motor neurons that control muscles. The nervous system directs an animal's movement, memory, decision making, and behavior.

To begin unraveling its mysteries, Sydney had to select an experimental organism. His previous favorite organism, *E. coli*, did not have a brain, so it was not a candidate. Brenner settled on the nematode roundworm, *C. elegans*, which offered many advantages. It is one of the simplest organisms that possess a brain. The adult nematode has only about 1,000 cells in the entire organism, and about 300 of these are neurons—the cells that power the nervous system. Furthermore, these nematodes are transparent, so the different cell types and their connections are readily observed in a microscope. Finally, these worms are small—you'd need to line up 25 of them to span 1 inch—and their generation time is only 3½ days, so they are cheap and easy to grow.

Sydney was so charismatic, so smart, and possessed of such a charming, gregarious personality that some of the most talented and adventurous young scientists of the time followed him into the world of worms. Andy Fire and Craig Mello were among these disciples. Andy studied worms as a postdoctoral fellow with Sydney in Cambridge, England, in the mid-1980s before returning to the United States and setting up his own lab at the Carnegie Institution. Craig was introduced to worms in 1982 at the University of Colorado Boulder by my colleague David Hirsh, another of Sydney's trainees.

Thanks to their worm experiments, the researchers in the Fire and Mello labs had discovered that double-stranded RNA had a powerful ability to interfere with gene expression. But it was at first unclear how this process worked (how could a double-stranded RNA possibly recognize a single-stranded RNA target?) or whether similar processes occurred naturally.

Follow-up experiments in laboratories around the world soon answered the first question. Scientists discovered a suite of previously undetected (or at least underappreciated) proteins that allow double-stranded RNA to silence gene expression. One of them was an enzyme, appropriately anointed *Dicer*, that chopped long double-stranded RNA into smaller pieces called *small interfering RNAs* (siRNAs). These were then bound and shepherded to their sites of action by another enzyme researchers dubbed *Argonaute*, after a 50-gun French ship of the line from 1708.

The moniker seemed appropriate, because the Argonaute protein carries some formidable artillery and, like a warship, it travels around looking for its target. In the protein's case, it's a strand of RNA that guides its search. Its target is a messenger RNA with a sequence complementary to one of the siRNA's strands, known as the "guide strand." Upon loading into Argonaute, the other, "passenger," strand of the siRNA is expelled—leaving the now-single guide strand free to base-pair with complemen-

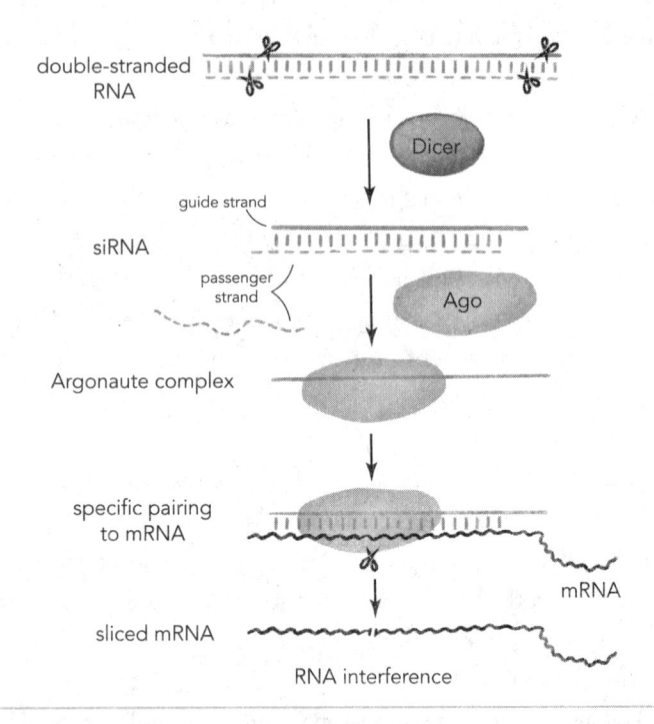

RNA interference starts with long double-stranded RNA, which is cleaved by the cellular enzyme Dicer to give 23-nucleotide siRNAs that are mostly base-paired. Upon binding to the Argonaute (abbreviated as *Ago*) protein, one of the two strands (passenger strand) is ejected, and the other (guide strand) travels around with Ago and binds to matching mRNA sequences. The Ago enzyme then cleaves, or "slices," the mRNA, thereby inactivating it.

tary sequences on the target mRNA. Then the artillery comes into play: The Argonaute protein is an enzyme, able to cleave and thereby inactivate the target mRNA held helplessly in place by base-pairing to one strand of the siRNA. It's like the siRNA is a missile-guidance system, directing the Argonaute warhead to its site of attack.

Presumably Dicer and Argonaute were not floating around in worms just waiting for researchers to come along and inject double-stranded RNA. RNA interference must have a normal biological function. But what was it? As it would turn out, the answer had already been found; it was just a matter of making the connection. Starting in 1993, developmental biologist Victor Ambros at Harvard and geneticist Gary Ruvkun at Massachusetts General Hospital had identified extremely small nematode RNAs, called *microRNAs*, which played an important role in the development from an embryo to a complete organism by switching off the production of various proteins at critical stages. These natural microRNAs were initially made as larger RNAs that base-paired within themselves to form long double-stranded segments, akin to the folding of each of the arms of the tRNA cloverleaf. They were then processed by Dicer and loaded onto Argonaute to inhibit the activity of natural mRNAs. That explained why worms were pre-equipped with the machinery needed to make use of artificially injected siRNAs.

Nature went to great lengths to evolve systems to manufacture complex proteins, so why would it need microRNAs and RNA interference to undo all that hard work? As an organism develops from an embryo to an adult, it must build different organs such as the brain, gut, skin, and reproductive organs. To embark on these different developmental trajectories, it's not enough only to make new types of proteins. Cells also need to quit making the old types of proteins. This is what microRNAs add to nature's toolkit: the ability to regulate the translation of specific mRNAs.

Each microRNA seeks out and inhibits not just one mRNA but a whole set of related mRNAs. The result is an enormously complex and intricate regulatory network. Instead of inhibiting gene activity, let's think of inhibiting traffic flow across the East River in New York City. Multiple bridges, including the famous Brooklyn Bridge, carry traffic into and out of Manhattan; these are analogous to the multiple genes whose activity needs to be reduced before an embryonic cell can become a brain cell. The traffic across each bridge can be inhibited by events occurring in Manhattan—road repairs, traffic accidents, or a sudden snowstorm. These events are akin to the actions of microRNAs on gene activity. Each bridge will be affected by these events to a different extent, depending on its location and other factors. The effects will be additive; combining a stalled truck and a sudden snowstorm will really shut things down. Similarly, with RNA interference, it's the combination of the number of microRNA binding sites—and the prevalence of those particular microRNAs—that tunes down the translation of an mRNA into protein.

This pathbreaking research was recognized by Nobel prizes to Fire and Mello in 2006 and to Ambros and Ruvkun in 2024. But in the late 1990s, these tiny RNAs were mostly known in worms. To unlock the medical potential of siRNA to silence genes, researchers had to see if the same magic worked in humans.

BEYOND THE WORM

Tom Tuschl was the man for the moment, a scientist talented and dedicated enough to help turn the therapeutic promise of RNA interference into lifesaving reality. We met in 1989, when he worked in my lab as an exchange student from Regensburg University in Bavaria. He struck me as industrious, intelligent—but I could not have guessed how important his later discoveries would turn out to be. By 1999, he was working on RNA interference in Phil Sharp's lab at MIT. They were the first to show that siRNA inhibits its target mRNAs not by some subtle action, but very directly—by slicing them up.

Tom then returned to Germany and quickly made key findings that set the stage for making medicine from RNA interference. One major question concerned the occurrence of microRNAs in humans. If they were present, then the machinery for using them—including the Argonaute "slicer" protein—must also be present. If this were the case, maybe one could commandeer that machinery by introducing a double-stranded RNA that targeted an mRNA involved in some disease, creating a powerful new drug.

First, though, scientists needed to confirm that microRNAs worked beyond the worm. To do this, Tom purified all the RNA from a wide variety of species, then separated the RNAs by size using gel electrophoresis, as Art Zaug had done during my lab's work on the *Tetrahymena* ribozyme. Using a razor blade, he cut out the portion of the gel that would contain any really small RNAs—about 21 to 23 base pairs—leaving the larger ribosomal, messenger, and transfer RNAs behind. In the end, he discovered dozens of previously overlooked microRNAs from fruit flies, fish, mice, and, most important, human cells. It now appeared that nature used RNAi to tune down genes in a broad swath of biology.

Given that each of these microRNAs is encoded in the genome and that the initial announcement of the human genome sequence occurred a year earlier in 2000, how had all these microRNAs flown under the radar? In everyday life, you start looking for your missing car keys under the lamppost because that's where the light is strongest, and science is not much different. Most scientific energy had gone into studying genes that coded for proteins, even though such protein-coding sequences collectively represented only 2 percent of the human genome. Surrounding those protein-coding islands was a veritable ocean of other DNA. So it was easy to overlook the tiny specks of microRNA genes.

In time, as many as 500 different microRNAs would be discovered in humans. They've been shown to contribute to multiple essential processes, including the proper development of arms and legs; the formation of heart muscle; the proper production of blood cells, particularly of immune cells; and placental development and pregnancy. When microRNAs are perturbed, they contribute to many diseases. For example, tumor cells generally stumble onto ways to decrease microRNA levels, thereby up-regulating genes that promote their growth. One microRNA, for instance, normally functions to keep genes that promote cell division under control. Cancer cells make less of this microRNA, thereby promoting inappropriate cell proliferation.

If changes in RNA interference can cause disease, might we be able to turn it against disease as well?

A GUIDING STAR IN ORION'S BELT

In the same year he found human microRNAs, Tom Tuschl discovered that it only took a small double-stranded RNA, the length of about 21 base pairs, to shut down gene expression. In other words, it wasn't necessary to treat cells with double-stranded RNA molecules that were hundreds of base pairs long—as scientists had been doing since they first heard about Andy Fire and Craig Mello's work—and then let Dicer cut that RNA down to size. Instead, they could just provide the short double-stranded RNA directly. Critically, because Tom had been trained by the pioneering nucleic acid chemist Fritz Eckstein in Göttingen, he was able to produce these RNAs by chemical synthesis. And if siRNAs could be chemically synthesized, they started to look a lot like drugs. Tom had laid the scientific groundwork for turning RNA into a therapeutic agent that could target mRNA arising from harmful genes.

In 2002, Tom Tuschl, Phil Sharp, and former Sharp lab colleagues Dave Bartel and Phil Zamore founded Alnylam Pharmaceuticals. Alnylam (or Alnilam) is the bright star in the belt of the constellation Orion, and just as Polaris points the way to the north, this star in Orion would hopefully aim the company at a whole new class of therapeutics.

What made siRNAs so attractive as potential therapeutic agents? Consider that development of any potential drug requires sorting out a host of questions, including: How specific is it for its target relative to healthy processes that might also be affected? What are its side effects, and can they be tolerated? What's the effective therapeutic dose? How often does it have to be taken? Traditional pharmaceuticals are small organic molecules, such as the aspirin we take for aches and pains or atorvastatin (Lipitor), which we take to lower our cholesterol levels. For such drugs, answering all the questions about safety and efficacy is a long and expensive research and development project, which must begin anew for every new molecule. There's a good chance that a drug will fail before surmounting all these hurdles. In theory, siRNA could vastly simplify this process. Certainly, the first time around it still presents numerous challenges: stabilizing the siRNA, figuring out how to deliver it to the relevant tissues in the body, and ensuring that it is safe and effective. But once these problems are solved for one application, then targeting a new disease could be as simple as changing the sequence of A, U, G, and C bases along the siRNA to match the new mRNA. The issues of stability, delivery, and safety would to a great extent already be "preapproved."

Alnylam chose to attack the problem of rare diseases, defined as diseases that affect fewer than 200,000 people in the United States. These are also known as "orphan diseases," because the patient population isn't big enough for pharmaceutical companies to deem it worthwhile to spend the billion dollars it takes to develop a drug and run it through human clinical trials. But in the aggregate, orphan diseases represent an enormous unmet medical need. More than 3,000 inherited disorders have been found to be caused by a mutation in a single gene, and some 25 million people in the United States alone are affected by one of them. While it might be impractical to develop 3,000 different drugs for 3,000 orphan diseases, could one develop a single siRNA drug and then tune its sequence to match 3,000 targets? Can all these orphans be given a supportive home?

The first challenge the Alnylam team faced in turning siRNA into an effective drug was working out how to deliver it to the disease-affected cells. RNA by itself is too unstable to be a good drug; it's easily degraded by the ribonucleases that are abundant in all human tissues because they break down RNA present in the food we eat or allow cells to switch their pattern of gene expression. Atop that, RNA can't pass through the protective membrane layer that shields cells from unwelcome intruders. So Alnylam scientists borrowed a trick that RNA-based viruses use all the time: they wrapped the RNA in an envelope, a greasy lipid shell, which then dissolves into the human cell membrane, letting the RNA inside. This packaging also protects the RNA from ribonucleases.

In its first clinical trial of an encapsulated siRNA, Alnylam targeted a disease called hereditary ATTR, or amyloidosis mediated by transthyretin (TTR). The TTR protein is made in the liver and normally acts as a transporter, helping maintain normal levels of thyroid hormone, vitamin A, and other molecules. But with a disease, it is sometimes less important how the normal protein behaves than how the mutant protein misbehaves. Inherited mutations in the TTR gene cause the TTR protein to misfold, accumulating as fibrils in the nerves and heart. Most of us have never heard of this disease, because it's rare: only about 50,000 patients worldwide. But for those 50,000 people, it's a disaster—they suffer from heart disease and neurological problems and often have trouble walking. Death usually occurs within about a decade of the original diagnosis.

Alnylam's siRNA drug accumulates in the liver, where the TTR mRNA is made, and it then prevents the mutant TTR protein from being produced. In 2018, the clinical trial of the siRNA was complete, and the news was good: the ATTR patients who had received the drug stabilized and in fact saw their walking ability improve, while those in the control group, who had received a placebo, continued to deteriorate.

But in drug development, you often solve one problem and another pops up. In the case of siRNA therapeutics, the nanoparticle-encased drug had to be delivered intravenously (IV) once a month, requiring patients to come to a hospital or an infusion center and sit for an hour while the medicine was delivered in a slow drip through a needle in the arm. It was expensive, tedious, and often painful for the patient. To avoid IV delivery, the Alnylam scientists found a way to deliver the siRNA by subcutaneous injection through fiddling with the double-stranded RNA. They added a kind of "handle" that would be grasped by a receptor on the surface of liver cells. While the trick was specific for liver cells, it allowed the siRNA to be administered by a quick shot in the arm, like a vaccination, rather than by IV.

Just as the scientists at Alnylam had hoped, their progress developing siRNA therapy for ATTR made it much easier to tackle the next diseases. Between 2018 and 2023, they had four additional liver treatments approved by the FDA, all for rare but very debilitating diseases. After taking 16 years to develop the first treatment, they have averaged one per year since. Of course, with 3,000 single-gene disorders to consider, there's still a long road ahead.

A GROWING THREAT

Alnylam has proved the viability of siRNA therapy to treat rare genetic diseases. But what about devastating diseases that are far too common? As medicine has advanced, fewer people die of infectious diseases. Even death from cancer is on the decline: from 2001 to 2020, cancer death rates in the United States decreased by more than a quarter, from 197 to 144 deaths per 100,000 people per year. But as people live longer, they are more likely to suffer from one of the terrible neurodegenerative diseases such as Alzheimer's, Parkinson's, or amyotrophic lateral sclerosis (ALS). The death rate from Alzheimer's and Parkinson's is increasing rapidly—more than doubling over the same 20-year period when cancer deaths have declined—and ALS is similarly on the rise. These diseases not only debilitate people but also destroy families, who are often overwhelmed with anger, fear, and grief as a loved one becomes unrecognizable.

RNA is directly implicated in all these diseases. So could some version of the siRNA technology be used to fight the rise of neurodegenerative diseases as well? For example, take ALS, also known as Lou Gehrig's disease after the baseball player who was afflicted by it. ALS is particularly devastating because it arises suddenly in seemingly healthy individuals and then progresses rapidly, attacking motor neurons. I've seen it strike twice in my wider circle of friends and colleagues. These individuals were at the height of productive lives when they progressively lost the ability to eat, to speak, to walk, and finally to breathe. In one case, paralysis and death occurred five years after the initial symptoms; in the other case it was one year.

Although many cases of ALS are *sporadic*, meaning that they arise where there is no family history of the disease, other cases run in families. Such *familial* cases are of special interest to biomedical scientists because they can lead to the identification of a genetic cause of the disease. The most common genetic cause of ALS involves a gene that goes by the very technical name C9orf72. The gene normally contains a few repeats of a certain genetic sequence: GGGGCC. But in ALS, mistakes in

DNA replication cause the repeat to undergo an enormous expansion—the gene contains thousands of repeats of the same sequence, GGGGCC. When this wonky DNA is then transcribed into RNA, the repeats are preserved. Scientists are still working to unravel all the problems caused by this unnatural RNA, but one major concern is the way this defective RNA attracts and holds on to proteins (including one called hnRNP H) that are required for proper RNA splicing. With so many of these RNA splicing proteins stuck to the RNA repeats, they have trouble doing their normal job, and the alternative splicing patterns on which neurons depend are perturbed. Ultimately, the neurons die, and patients' bodies lose their ability to transmit signals from the central nervous system to peripheral muscles.

With RNAi therapy, could scientists one day chop up the pathogenic RNA that contributes to ALS, halting its progress—or even prevent it from forming in the first place? This is speculative, to be sure. For example, delivering siRNA to motor neurons will be a much greater challenge than delivering it to the liver. You can access an organ such as the liver by putting a drug into the bloodstream. But reaching the brain is a trickier proposition. That is because of a natural defense system, called the blood-brain barrier, a wall of tightly packed cells that evolved to keep toxins and other harmful agents away from brain tissue. This barrier would filter out any RNA-based drug before it reached the brain, meaning it would have to be delivered through another process, such as injecting it into the fluid that surrounds the spinal cord in the backbone—an invasive and costly procedure. In addition, because chopping up the pathogenic RNA would not restore the normal gene function, siRNA therapy might fall short of being therapeutic. But given the scientific promise of this technique and the growing medical need, researchers are not giving up on aiming siRNA therapeutics at the brain. In the fight against ALS and other neurodegenerative diseases, we need all the firepower we can muster.

Alzheimer's dementia is another terrible neurodegenerative disease that could potentially be treated with siRNA. In 2021, more than 6 million people in the United States alone were living with Alzheimer's disease, and the number rises each year as the population ages. Two types of protein aggregates, called amyloid plaques and tau tangles, accumulate in the brains of these patients and are thought to inhibit proper neuron function. In the first case, a protein called amyloid precursor protein is cut up by brain enzymes, resulting in a protein by-product called beta-amyloid. This material accumulates between neurons the same way that plaque does between teeth. In the second case, a protein called tau can accumulate not around neurons but inside them, resulting in tangles that make the receptors go haywire. Like all human proteins, amyloid precursor protein and tau are each encoded by an mRNA that directs their synthesis. So it seems plausible that an siRNA that cleaved one or both of these mRNAs, thereby reducing the amount of the corresponding proteins, could be therapeutic.

In 2022, Alnylam announced a new program in collaboration with Regeneron, the biotech company best known for its antibody treatment against Covid-19, to tackle Alzheimer's disease. More specifically, they are developing siRNAs that target the mRNA for amyloid precursor protein. They expect that reduction of the levels of this protein will correspondingly decrease the formation of beta-amyloid plaques. By replacing the previous liver-targeting "handle" on the siRNA with a new handle, they have already managed to silence amyloid precursor protein in the mouse central nervous system safely and effectively.

The road from demonstrating that a therapy is effective in mice to establishing an effective treatment for humans is a long one, littered with many potholes, but we should all hope for the best. After all, the stakes are so high: reversing neurodegenerative disease is now arguably the most challenging unmet medical need of humankind.

From the lowly worm came therapeutic siRNAs. It's certainly a remarkable story, but fortunately, it's not at all unique. The biggest breakthroughs in biomedicine almost always come from fundamental research that's being done to understand how nature works, without any medical application in mind. Andy Fire and Craig Mello wanted to study genes that control the behavior of a tiny transparent worm, trusting that what works for the worm will be applicable to other multicellular organisms, including humans. To knock down the production of particular gene products, they wanted to improve their toolkit, and antisense RNAs seemed promising. But by performing creative experiments, coupled with a good dose of serendipity, they found that the winning RNA was not a single strand but a double-stranded version. That dramatic discovery opened the door for an entire new field of research—microRNAs that regulate gene pathways in all complex organisms, including humans—and an entirely new class of therapeutic agents.

We now understand some of the damage that RNA can do when it malfunctions. But good RNAs that have gone bad, such as the kinds that cause or contribute to neurodegenerative diseases, are not the only ones to worry about. Some RNAs are, at least from our perspective, born to be bad. As we'll soon learn, many of the viruses that cause enormous pandemics are powered entirely by RNA. But, although RNA has a dark side, understanding how it works also suggests ways to fight it using its own tricks.

Chapter 9

PRECISE PARASITES, SLOPPY COPIES

It was 1935, and biochemist Wendell Stanley was busy tending his Turkish tobacco plants in the greenhouse at the Rockefeller Institute for Medical Research in Princeton, New Jersey. When his seedlings had grown to a height of 3 inches, he took a bandage gauze pad, dipped it into his stock of tobacco mosaic virus, and rubbed it on the leaves. This virus, the bane of the tobacco industry, got its name from the tiled pattern of blotches it produces on the infected leaves. Scientists in the late nineteenth century had discovered that some infectious agents are so small, they can pass through a filter that retains bacteria; these agents were then named viruses. Forty years later, scientists still had no idea what viruses were made of, let alone how they caused infections. This was the knowledge gap that Stanley was aiming to fill.

Three weeks after Stanley rubbed the tobacco virus on the leaves, the infection was in full swing. He cut the plants, froze them, then pushed the frozen plants through a meat grinder—not a very sophisticated scientific instrument, but an effective one. After allowing the pulp to thaw, he pressed out the juice, thick with the object of his fascination: the viral particles. He could see them in the electron microscope—beautiful little rods, shorter than an *E. coli* bacterium and much, much thinner. No wonder the virus could pass through a filter that trapped bacteria.

Stanley's solution of tobacco mosaic virus particles was so concentrated that he was able to crystallize the virus. As we saw with James Sumner and his urease protein crystals, crystallization was a well-accepted method for obtaining a substance in pure form: the crystal contains only the molecules of interest, and any contaminants are left behind. When Stanley analyzed the composition of his virus crystals, he found that they were pure protein—or, *almost* pure protein. The crystals were 94 percent protein; there was also a pesky 6 percent of RNA.

Stanley's discovery was deemed remarkable. A virus, which had the lifelike properties of reproduction and mutation, had the simple chemical composition of protein. When Stanley gave his lecture accepting his share of the 1946 Nobel Prize in Chemistry, which celebrated the power of proteins,* RNA was relegated to a single footnote. Yet in the intervening decades, the significance of that footnote would grow and grow. While Stanley saw RNA as an unimportant detail, we now see it as the key to understanding, and fighting, many of nature's most fearsome viruses.

FREELOADERS

Viruses are inevitable. Whenever there is an organized biological system—a cell, an organism, a community—entities that take advantage of that system to benefit themselves, without adding any value, are certain to arise. We call them parasites. They are inevitable because it is vastly simpler and easier to be a parasite than to be a fully functional organism. The very chemical principles and environmental conditions that give rise to organisms thus give rise to parasites, too.

Because they are inevitable, the number of viruses on Earth is vast.

---

*Stanley shared the Nobel Prize with James Sumner, who had crystallized the enzyme urease and thereby established that "all enzymes are proteins," and John Northrop, who helped show the generality of Sumner's work.

At 1×10^{31}, that number is 10 billion times larger than the number of stars in the known universe. Lucky for us, most of these viruses are phages that infect only bacteria. Their diversity is so great that when undergraduates in our molecular biology course in Boulder purify their own bacteriophage from soil or a local landfill or the lion cage at the zoo, each phage is inevitably a never-before-seen entity.

Every virus needs to replicate itself; every virus needs a set of genes to carry out its infectious cycle. The question that Stanley and other scientists needed to answer to unlock the mystery of viruses was how that genetic information was stored. Was the information really embedded in the protein? Stanley won the Nobel Prize in 1946, two years after Oswald Avery of the Rockefeller Institute in New York City announced that DNA is the "transforming principle," composing the genes in bacteria that cause pneumonia. Yet the idea that the genetic material might reside not in nucleic acids but in more-complex protein molecules still hung on. And in his Nobel Lecture, Stanley punted on the question. He never placed a bet on the chemical nature of the virus's heritable material. Was it the protein, which clearly made up the bulk of the virus—and was the subject of his Nobel Prize? Or might it be that relatively small amount of nucleic acid?

In any case, Stanley underestimated RNA—and, as we have learned by now, one should *never* underestimate RNA. But when he moved to the University of California, Berkeley, in 1948 and recruited the team for his new Virus Laboratory, he hired someone who would give RNA the attention it deserved.

Heinz Fraenkel-Conrat's path to Berkeley was circuitous. He was born in the ancient city of Wroclaw, in what is now Poland, and obtained a medical degree there in 1933. Seeing the rise of Nazism in Germany, he wisely moved to Edinburgh for his PhD studies and then emigrated to the United States. His biochemist brother-in-law, Karl Slotta, had moved from Poland to São Paulo, Brazil, where he was following up on his discovery of the hormone progesterone, which would lead to birth-control pills. Fraenkel-Conrat visited Slotta in Brazil and stayed on; they ended up studying the venom of the South American rattlesnake, using it to purify the first neurotoxin. In 1952, Stanley recruited Fraenkel-Conrat to Berkeley to work in his new Virus Laboratory.

In Berkeley, Fraenkel-Conrat was intrigued by the smidgen of RNA in tobacco mosaic virus (TMV). He built on the discovery of two German scientists, who in 1956 showed that purified TMV RNA scratched onto tobacco leaves would cause infection with TMV. No protein seemed to be required—a scratch of the leaves was enough to allow the RNA to enter. These were indeed powerful results supporting RNA as the TMV genetic material—unless, perhaps, there was an undetectable bit of TMV protein coming along with the RNA and causing the infection.

To rigorously test the idea that the viral RNA is TMV's genetic material, Fraenkel-Conrat purified RNA from a strain of TMV that produced only small local blotches on the leaves instead of a systemic infection. He then assembled that RNA with the protein from the fully active TMV strain. When this virus was scratched onto the plant leaves, it resulted in an infection of the small-blotch type. Conversely, when he assembled the opposite mixture—containing RNA from the fully active strain and protein from the small-blotch strain—the fully active strain won out. The RNA, not the protein, determined the outcome of the infection, showing that RNA is clearly the genetic material of the virus.

WHO NEEDS DNA, ANYWAY?

Viruses come in two major flavors. Some viruses, such as those that cause chicken pox and smallpox, encode their genetic material in DNA, just like plants, animals, and every other living thing on Earth. But many of the worst viruses have genes made of RNA and don't bother with DNA at all. These RNA viruses include not only those that cause plant diseases such as tobacco mosaic but also those that cause human diseases such as influenza, measles, mumps, polio, Zika, Ebola, and Covid-19, to name just a few.

Though these RNA viruses can do without DNA, the converse is not true. DNA viruses still need RNA. Just like more complex organisms,

DNA viruses transcribe their DNA into mRNAs that then encode viral proteins. This makes RNA the common denominator of all viruses.

How ancient are RNA viruses? We've already considered that the first self-reproducing system on Earth could have consisted of RNA, because RNA can be both an informational molecule and a biocatalyst that replicates the information. That is the RNA world hypothesis for the origin of life. I bet that about a day after the first RNA self-replication system took off, a small scrap of parasitic RNA was already hitchhiking along for the ride, being replicated but adding no value to the system. And each time a new organism arose on the planet, its viruses came soon thereafter. What's more, the fact that viruses can mutate to change their host range—from animals to humans, for example, a process known as zoonosis—means that there's always a large reservoir of potential new viruses at the ready.

Over millions of years of evolution, all RNA viruses had to surmount the same hurdles: how to infiltrate host cells, how to make the proteins they need, how to replicate their RNA genomes, and how to package themselves into infectious offspring. Each virus solved these problems in different ways. In the case of SARS-CoV-2, for instance, the virus has evolved its signature Spike protein, which fits into a receptor called ACE2 on the surface of the cells in our nasal passage and lungs like a power plug fits into an electrical socket. Once the virus locks onto the surface of the cell, it has to sneak itself past the cellular defenses. Not a problem. Because the virus is covered by a *lipid envelope* and the cell has a similar lipid envelope, the two can simply fuse. It's like the surface of a bowl of chicken noodle soup. There are flat islands of fat floating on the surface, surrounded by broth, and when two of the fat islands encounter each other, they fuse to form one larger fat island. Now the virus is inside the cell and free to do its worst.

MAKING MORE OF YOURSELF

How does an RNA virus copy its RNA genome, anyway? Well, it depends. There are two major classes of RNA viruses: so-called *positive (+) strand* viruses such as SARS-CoV-2, and *negative (−) strand* viruses such as the influenza (flu) viruses. The (+) strand viruses replicate by first making a (−) strand, which is then used to make more infectious (+) strands. Think again of how you would go about making a plaster cast of a garden gnome: you first make a mold, which is an inverse replica of the gnome, and then you pour plaster into the mold and cast as many replicas of the original gnome as you wish. It's the same with the (+) strand viruses: once a complementary (−) strand is synthesized, it can be used over and over again to make (+) strands. The other key feature of (+) strand RNA viruses is that the viral RNA that infects a cell also serves as a messenger RNA. Once this mRNA enters the host-cell cytoplasm, it finds human ribosomes, which, not knowing that they're doing anything wrong, go about making the proteins that the virus needs for its infectious cycle. These proteins include the viral RNA polymerase that replicates the viral RNA. And they include the *capsid* protein and the Spike protein that coat newly made particles and allow the virus to be infectious.

Tobacco mosaic virus is a member of this positive club. Other (+) strand RNA viruses that infect humans include poliovirus, dengue virus, hepatitis A and C, and rhinovirus, which causes the common cold. Rubella, or German measles, is caused by a (+) strand RNA virus that was one of the banes of childhood until it was largely suppressed by measles, mumps, and rubella (MMR) vaccination.

Conversely, so-called negative (−) strand viruses enter the host not as an mRNA ready to code but as its complement. In other words, they come in not as a garden gnome but as the mold for a gnome. These viruses bring along their own copying enzyme, and once they enter a cell, the enzyme goes to work copying the (−) strands into (+) strands that serve as mRNAs. These viral mRNAs once again hijack the host cell's ribosomes to produce their poison proteins. All the flu (influenza) viruses are (−) strand, as are respiratory syncytial virus (RSV), rabies virus, and Ebola virus. Mumps and measles viruses also belong to this group, so the MMR vaccine protects against two (−) strand viruses and one (+) strand RNA virus.

How many proteins does a viral mRNA encode? The number varies quite a bit, but it's usually not very many. Viruses are the ultimate efficient parasites, doing as little work as possible and tricking their host into bearing the load of most of their infectious cycle. TMV is remarkably efficient: Its RNA is only 6,300 bases and encodes four proteins, two of which handle RNA replication, one of which facilitates cell-to-cell transfer of the virus within the plant, and the last of which forms the cylindrical coat, or capsid, of the virus that sequesters the RNA in its central cavity. The polio and flu virus genomes encode 10 and 17 proteins, respectively. SARS-CoV-2 is a real monster (in more than one way), with a genome encoding 29 proteins. While that is huge for a virus, it's only a tiny fraction of what it takes to make a real organism. *E. coli*, for example, encodes about 4,000 proteins, while a human being encodes about 20,000 proteins.

A FEW TYPOS HERE AND THERE

My daughters text me a lot, and their fast fingers frequently foment errors: "Sounds food. Thank you!" followed a few seconds later by "food = good," or "I'm going to bake the kids up at 3" followed by "*wake." Occasionally the text has several typos in key spots, and I have absolutely no idea what the message means.

Viral RNA replication is the same. A few errors tend to be tolerated or can even be advantageous, but too many errors and the virus can't survive. The polymerases that copy the RNA make mistakes about once every 10,000 bases. That may not seem very error-prone; after all, we make mistakes in our everyday chores much more frequently than once every 10,000 times. But because viral genomes are around 10,000 bases long, such an error rate means that every time the RNA is replicated, a mistake has usually been made somewhere. Most of these are what scientists call base-substitution errors (such as putting in an A instead of a G), which often lead to changes in one amino acid in one of the viral proteins. Such a change might be neutral; it might hinder the virus's ability to go about its business; or, occasionally, it might improve the virus's fitness—for instance, if it helps the virus bind to a target cell more avidly, replicate more rapidly, resist antiviral drugs, or evade antibodies.

An irascible University of Illinois scientist named Sol Spiegelman was one of the first to show directly how viruses gain value from their mistakes. He was a refreshing presence in a sometimes-staid field—the rare biochemist who used the word "Biblical" to spice up normally dry scientific papers. In 1961, Spiegelman became intrigued by the question of how RNA phage replicated their genomes once they got inside bacteria. Such replication was key to the virus's survival, and yet at the time scientists had only a foggy idea of how it worked.

To answer the question, Spiegelman needed to get his hands on a phage RNA polymerase, the enzyme that RNA viruses use to copy themselves. He found that a phage called Q-beta makes a polymerase that is a well-behaved, stable enzyme, easy to purify. Q-beta is a (+) strand virus, and in test-tube experiments, Spiegelman saw how its enzyme used the RNA that came in with the virus as a template, made a complementary copy of it, and used that complement to reel off multiple phage RNAs. Plus makes minus makes plus.

For his most insightful experiments, Spiegelman made the bold move of dispensing with the bacterium and even with the virus. He started simply mixing the Q-beta RNA with its polymerase and watching it replicate and evolve in the laboratory—in a period of just a day. His experiments would help us understand how errors in replication produce variant viruses that acquire new capabilities.

One of the first evolutionary experiments Spiegelman conducted addressed the question: "What will happen to the RNA molecules if the only demand made on them is the Biblical injunction, *multiply*, with the biological proviso that they do so as rapidly as possible?" To achieve this, Spiegelman conducted what's called a "serial transfer" experiment. He set up a row of test tubes, each holding a simple salt solution containing RNA nucleotides, the building blocks for replication of new RNA. He put a drop of Q-beta RNA plus polymerase in the first tube. Twenty minutes later, the tube was dense with replicated RNA. He took out one drop from that and used it to "seed" tube no. 2. After a few rounds in which he let replication proceed for 20 minutes, he raised the stakes by decreasing the time between transfers to 15 minutes, then 10, then 5. In this way, he put pressure on his system so that, each time, the most rapidly replicating molecules would win out, eventually taking over the population.

After a day of evolution, Spiegelman looked at what was in the final tube. His original 3,300-nucleotide viral RNA had whittled itself down into a "little monster" containing just a few hundred nucleotides. He realized that the Q-beta polymerase had been making mistakes, occasionally jumping over part of its RNA template. Under conditions in which fast replication was rewarded, fewer bases to copy had offered a selective advantage, and the little monster won the contest.

Spiegelman tested other selective pressures. When he added a pinch of ribonuclease to the replicating phage RNA, most of the RNA was degraded and lost, as you'd expect—RNA really hates ribonuclease. But a rare RNA molecule that happened to have mutations at the sites where the ribonuclease preferred to cut was somewhat protected. After multiple rounds of replication, a ribonuclease-resistant mutant had arisen and was happily replicating in the presence of ribonuclease.

This phage RNA evolution presages what we've seen recently with the SARS-CoV-2 viral RNA. As SARS-CoV-2 tore through the global population, it mutated countless times, and some of those mutations turned out to give it an edge. Take the Omicron variant, which was first reported to the World Health Organization in November 2021, nearly two years after the first cases of Covid-19 were identified. Omicron has 35 mutations in the Spike protein compared to the original Wuhan strain of the virus, each of these mutations causing a single amino acid change. Located in the part of Spike that binds to the receptor on the outside of the human cell, these mutated amino acids increase its ability to lock onto the surface, presumably explaining why Omicron is so much more infectious than earlier variants. At the same time, these mutations make the virus better at fending off the antibodies that were produced against previous versions of the Spike protein, making therapeutic antibody treatments and vaccination less effective.

It's not that the virus is *trying* to evade the antibodies. Rather, its replicase is making mistakes and thereby inadvertently testing new variants all the time. Those mutant viruses that just happen to evade the human immune response are able to live long and prosper.

PLEASE ENVELOP ME, AGAIN

Astronauts orbit Earth in the confines of a space capsule, which serves two major functions: protecting them from the perils of outer space and guiding them back to Earth when their mission is complete. Like an astronaut, viral RNA cannot move around as naked RNA but needs to be enclosed in a capsid. The capsid protects the RNA from the perils of human tissues—such as ribonucleases—and it guides the viral RNA to its target cell. Having a capsid is so important that the viral RNA uses part of its limited genome to code for a protein or proteins that assemble with the RNA to form the capsid.

As Sol Spiegelman's experiments demonstrated, viruses are under pressure to keep their genomes small to allow them to replicate quickly, so each gene is precious. As a result, viruses make their capsid with a minimal number of protein building blocks. TMV does it with just a single type of protein building block, with each protein molecule locking into the one before it and the one underneath it. The proteins are made to assemble in a curved array, so the end result is a cylindrical tube with a hole in the center that accommodates the RNA. The construction process is like building a wall from identical wedge-shaped Lego bricks, snapping them together so that the wall circles around to form a tube.

Phage Q-beta makes a capsid of a very different shape. The ancient Greeks explored geometry and came up with the Platonic solids—3D structures assembled only from identical shapes. One of the simplest is the icosahedron, an almost spherical box formed of 20 triangles. But long before the ancient Greeks, phage Q-beta was assembling its little home

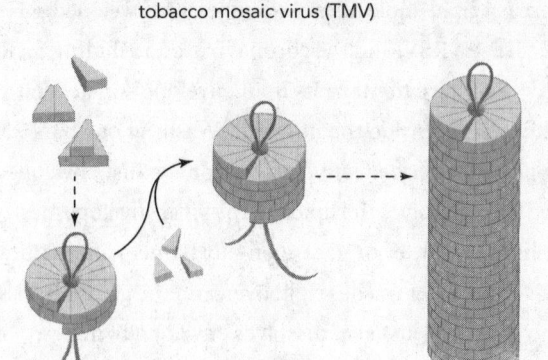

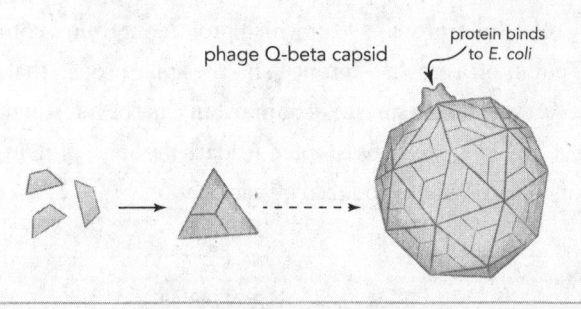

Each virus makes its own distinctive capsid to encapsulate its RNA genome, protecting it and helping it infect cells. TMV RNA encodes a single wedge-shaped protein that assembles with the RNA to form a long cylindrical tube. Phage Q-beta RNA encodes both a capsid protein, many copies of which assemble into an icosahedral shell with the RNA inside, and a second protein that binds to *E. coli* and facilitates viral RNA entry.

in the shape of a near-perfect icosahedron. One hundred seventy-eight copies of a single phage-encoded protein self-assemble to form most of the icosahedron, the little box that holds the viral RNA. Then a single copy of a second protein seals off the box. This protein also binds to hairlike projections on *E. coli*, helping the virus recognize and enter its bacterial prey.

For some RNA viruses, the capsid provides enough of a "space capsule" to protect the RNA and deliver it to its destination. In other viruses, however, the capsid is surrounded by yet another layer—an envelope made of greasy lipid molecules. Examples of enveloped RNA viruses include flu viruses, RSV, and the coronaviruses, including SARS-CoV-2. The virus doesn't have to make its lipid envelope—instead, it steals it as it's assembling itself inside the host cell. Washing our hands with soap and water provides very effective protection against enveloped viruses, because the soap dissolves the lipids in the viral envelope, destroying the virus. Washing off butter or grease on our hands with plain water isn't very effective; the water washes right over, and the grease stays stuck. But soap dissolves grease, just as it dissolves enveloped viruses.

As it puts on this new lipid overcoat, an enveloped virus decorates the overcoat with one or a few proteins of its own making—for example, the SARS-CoV-2 Spike protein, 90 of which project out from a coronavirus like the points on a crown ("corona"). It's the Spike protein that binds to a specific receptor on the surface of human lung, nasal, intestinal, skin, or brain cells, allowing the virus to enter. And it's the Spike protein that's the target of the antibodies produced by vaccination.

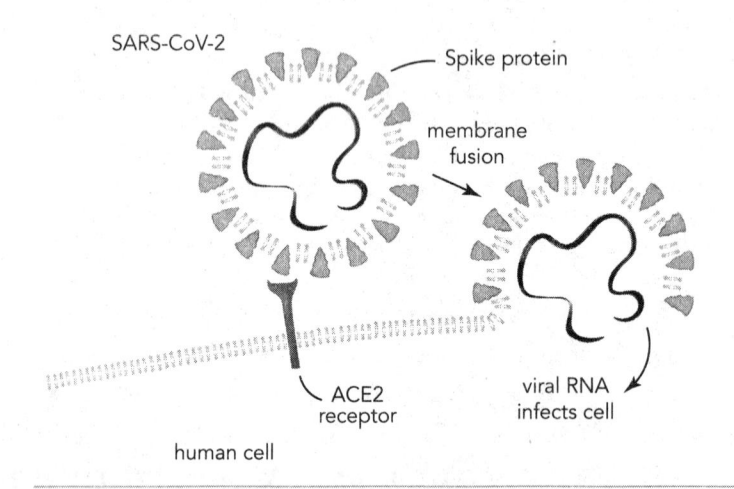

Enveloped viruses such as SARS-CoV-2 infect human cells by binding to receptors on the cell surface and then fusing with the human cell membrane, allowing the viral RNA (dark strand) to enter. The ACE2 receptor is anchored in the cell membrane, composed of lipids. Each lipid has a negatively charged head group and two fatty "tails" that interact with each other to form a double layer.

The mature virus particles exit the cell by hitchhiking on a pathway—*exocytosis*—that the cell has developed to export some of its own proteins. Altogether, a single SARS-CoV-2 virus that enters a cell produces about 600 progeny in about 8 hours. If each of these goes on to infect another cell, then the single virus produces 360,000 viruses in 16 hours and 216 million viruses in 24 hours. No wonder we can go from feeling completely fine to being laid flat by the viral infection so quickly.

We talk about viruses as a real scourge. They inconvenience or incapacitate us, our families, and our friends, interfering with the productive tempo of our lives. Sometimes they even kill some of us. Nonetheless, it's hard not to admire how efficient they are. It's absolutely remarkable that they can turn the world on its head with just a few dozen genes. Of course, they're completely dependent on a clueless host to pitch in and provide most of what they need for their infectious cycle. They are consummate exploiters.

Adding to their list of accomplishments, viruses are so very adaptable. They make enough errors in copying their RNA that most every virus is subtly different from its brother and sister. So when their surroundings change—for example, when they're beset by antibodies in our immune systems or by antiviral drugs—there's usually some virus in the population that just happens to have a solution to the new challenge.

It's only by understanding what viruses are made of and how they work that we're able to fight them effectively. As the Covid-19 pandemic taught us, a good way to fight an RNA-based virus is with an RNA-based vaccine. With human ingenuity, we've been able to turn RNA's genius against itself.

Chapter 10

RNA VERSUS RNA

After developing the first effective vaccine against polio in the 1950s, Dr. Jonas Salk was given the chance to build the research center of his dreams on 27 oceanfront acres in La Jolla, California. Salk asked the architect Louis Kahn to create something "worthy of a visit by Picasso." The resulting assemblage of teak and concrete blocks is today famous as both an architectural icon and a bastion of cutting-edge science. But very few people realize that the Salk Institute—named for the man whose vaccine saved the world from one pandemic—is also the birthplace of the idea of mRNA vaccines, which many years later would help bring another pandemic under control. This fact remains obscure because the journey from revolutionary idea to lifesaving reality had more twists and turns than the rocky coastline upon which the Salk is perched.

In 1989, Bob Malone was a graduate student at the Salk working in the laboratory of Inder Verma, an expert in using viruses to transport genes into human cells. Such technology was key to the nascent field of *gene therapy*—which uses DNA to treat or prevent a disease, most commonly by giving a person a new copy of a healthy gene to compensate for a defective gene. Obvious targets for this approach included genetic diseases such as sickle cell disease, muscular dystrophy, and cystic fibrosis, where the mutated gene was well known. Because gene therapy had the potential to provide a permanent cure to such disorders, it was a hot topic in the universities and companies around San Diego at the time.

In addition to gene therapy, scientists at this time were also working on a related idea: *DNA vaccines.* The two techniques share a fundamental concept: instead of introducing a useful protein molecule into a person, one can take a shortcut by introducing the gene for that protein, then relying on the human cells to copy the DNA into mRNA and then into protein. But unlike gene therapy, where the goal was to make a permanent change in the human genome, in the case of DNA vaccines even *temporary* expression of a protein—a viral or bacterial protein—could be enough to train the human immune system to be on the lookout for an unwanted invader.

In the 1980s, DNA therapeutics was an exciting possibility—but, in the lab at the Salk, Bob Malone saw a major potential drawback. When the DNA was injected into people, there was no telling where exactly in the patient's own genome it would land. Consider, for comparison, a wrench—inherently a useful tool, good for securing a nut on a bolt. Now imagine tossing that wrench at random into an automobile. It might land in an innocuous spot—the floor, a seat cushion, the trunk, the glove compartment, or the dashboard. But it could also land somewhere in the engine, a wheel well, a coil spring, or the driveshaft, where it could impede the car's function. And if the wrench wedged itself under the brake pedal or on top of the accelerator pedal, the car might drive out of control. A similar concern applied to DNA therapeutics. If the DNA encoding a foreign gene landed randomly in a patient's genome, it might settle in a nonessential part of a chromosome, doing no harm. But if by bad luck it interrupted a healthy gene or activated the expression of a nearby growth-promoting gene, it could cause a disease such as cancer. Indeed, some years later, a case made headlines when a child given gene therapy for severe combined immunodeficiency, the "bubble boy disease," developed leukemia after the therapeutic DNA happened to jump into a chromosome near a cell growth-promoting gene and turned on its activity.

Malone and Verma saw that it might be possible to get around this problem by using mRNA instead of DNA to instruct the body to make a therapeutic protein. mRNA could not be incorporated into a patient's genomic DNA, avoiding the possibility of a permanent and unwelcome change. It would take 30 years, but a descendant of this once-esoteric idea eventually became a household name: mRNA vaccines.

Even by conservative estimates, mRNA vaccines have saved millions of lives in the fight against Covid-19. They are now being developed not only for use against other viruses (from RSV to the common cold) but also against cancer. While the future of mRNA vaccines appears bright and perhaps even revolutionary, the history of their development remains murky and poorly understood. And as with any public health issue, that lack of clear information has led to confusion and has created fertile ground for conspiracy theorists.

I remember when mRNA vaccines were first making headlines back in the spring of 2020. News reporters and people on social media were talking about mRNA as if it were some foreign substance, a new drug. Many did not appreciate that, while mRNA was indeed being newly configured for use as a vaccine, it was also a natural and essential part of every cell in our body and every cell in every other organism on Earth. This lack of understanding of mRNA helped fuel apprehension that the mRNA vaccines were somehow categorically dangerous.

Yet it was not only ignorance about the nature of mRNA that contributed to suspicion of the vaccines. Another factor was the blazing speed with which these vaccines appeared. It staggers the mind that safe and effective mRNA vaccines were conceived, manufactured, tested, and approved for emergency use all within a year, compared to the six to eight years normally required for vaccine development.

How did scientists do it so quickly? The short answer is, they didn't. While the Covid-19 vaccine did appear in record time, it was built on decades of scientific breakthroughs. It might help to think of the vaccine as a jigsaw puzzle. It was impressive to see how quickly the puzzle was solved, but all the pieces were already lying there on the table when the pandemic struck. The challenge was figuring out how to fit them together. And the bonus for solving the puzzle was realizing that the knowledge we gained could be repurposed to develop mRNA vaccines for new viruses and other life-threatening illnesses.

THE FIRST PUZZLE PIECE

The first thing needed to make an mRNA vaccine—especially one intended to immunize a sizeable portion of the world's 8 billion people—is the ability to synthesize mRNA on demand, ideally by the truckload. It was an industrious biochemist named Bill Studier who found this first essential piece of the puzzle, back in the early 1980s.

Born in 1936, Studier grew up in Waverly, a small town in Iowa. He worked his way into Yale, then Caltech for a PhD, then Stanford for postdoctoral research. By 1964, he was at Brookhaven National Laboratory on Long Island, New York, heading up a research group investigating bacteriophage T7, a virus that infects *E. coli* bacteria.

Brookhaven National Lab inherited the facility that had been Army Camp Upton, which was decommissioned after World War II and turned into a research center dedicated to developing peaceful uses of atomic energy. The lab also had a biology department, where Studier reveled in the freedom to engage in curiosity-driven research, free of any requirement for commercialization or potential medical application. Phage T7 certainly fit the bill. How could a DNA virus that infected bacteria possibly have any medical relevance?

Studier was captivated by the incredible efficiency with which T7 commandeered the bacterium for its own purposes, turning the hapless cell into a factory dedicated to making more phage. He learned that the phage first hijacked *E. coli*'s own DNA-to-RNA copying machine, its RNA polymerase, and used it to copy the gene for the phage's RNA polymerase. Once the phage RNA polymerase appeared, it didn't waste any effort working on *E. coli* genes—it was custom-made to copy only phage genes into mRNAs for phage proteins. The phage polymerase had such extraordinary specificity because it required a certain sequence of 17 DNA base pairs—termed an "initiation site"—that sat at the beginning of each phage gene. This allowed the polymerase to completely ignore all the neighboring bacterial genes, which lacked this initiation site. In short, T7's RNA polymerase was a superefficient RNA-synthesizing machine.

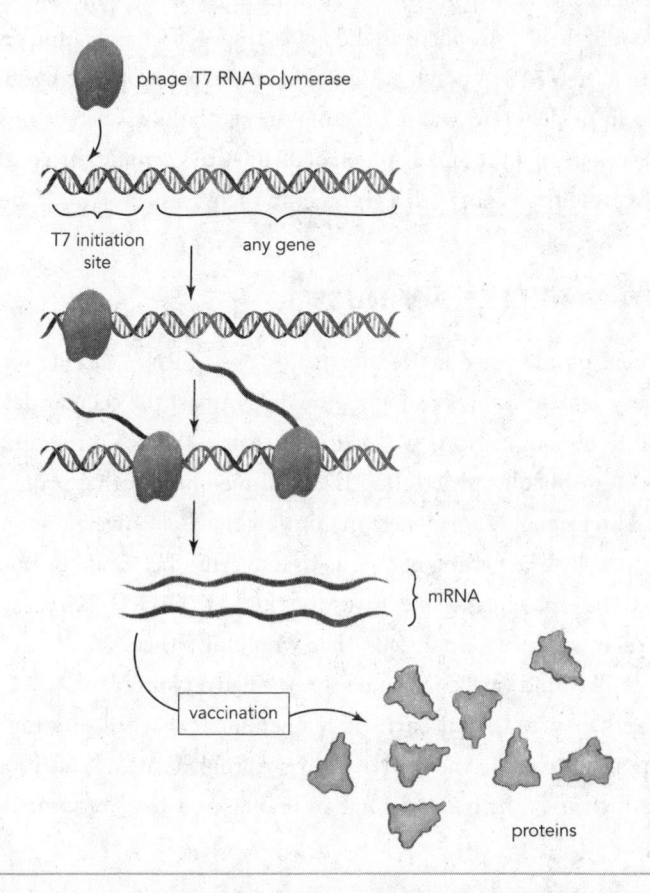

Phage T7 RNA polymerase can be used to transcribe essentially any RNA of interest. The scientist adds a 17-base-pair T7 initiation site preceding the gene to be copied. The polymerases then assemble nucleotide building blocks into mRNA (two polymerases shown here are transcribing from left to right). The last step shown here requires ribosomes to read the mRNA code and produce proteins. For example, a vaccinated person's cellular ribosomes use the injected mRNA to make the Spike protein.

As early as 1981, Studier predicted that the T7 RNA polymerase might be repurposed for producing the mRNA for any protein of interest. Shortly thereafter, he and his colleague John Dunn made his prediction a reality. They managed to isolate the gene of phage T7 that encoded the T7 RNA polymerase. When they put this gene into *E. coli*, the whole bacterial cell more or less filled up with T7 RNA polymerase, which they could then easily purify. It did not escape Studier and Dunn's attention that the T7 RNA polymerase would be useful for producing specific RNAs in the laboratory and directing the synthesis of specific proteins inside cells.* That the Spike protein of SARS-CoV-2 would turn out to be such a protein was not even a glimmer in their eyes.

THINKING LIKE A VIRUS

The next puzzle piece needed for making an mRNA vaccine was the delivery vehicle, a way of slipping the mRNA past the cellular defenses and into human cells. As we've already seen with siRNA therapeutics, RNA has difficulty penetrating the lipid membranes that protect our cells. Phil Felgner was a pioneer in thinking that lipid envelopes were not just a problem for getting nucleic acids into cells; they could be the solution to the problem as well. He was convinced that if RNA viruses could find their way into human cells, then wrapping a therapeutic mRNA in a virus-like lipid envelope could solve the delivery problem.

Perhaps it was Phil's artistic temperament that led him to take a creative approach to the mRNA delivery problem. Before settling on a scientific career, he had lived in San Francisco, where he studied clas-

---

*Multiple scientists contributed to development of bacteriophage polymerases as tools for making RNA. Other key contributors included Mike Chamberlin at UC Berkeley (purified the bacteriophage T7 and SP6 RNA polymerases), Doug Melton, Tom Maniatis, and Michael Green at Harvard (developed the bacteriophage SP6 system), Stan Tabor and Charles Richardson at Harvard Medical School (for the T7 system), and Olke Uhlenbeck at the University of Colorado (adapted the T7 system to use synthetic DNA templates for much more rapid production of small RNAs).

sical Spanish guitar and performed in coffeehouses. After completing his postdoctoral research in Virginia, he moved back to the Bay Area, took a position at Syntex Corporation, and began developing his lipid delivery vehicles for mRNA.

It's fortunate that Phil was brave enough to stake his future on lipids. Most RNA and DNA scientists, in contrast, generally try to avoid even thinking much about these fatty and oily molecules. RNA and DNA are well behaved—easy to isolate in pure form and easy to dissolve in water. Lipids, on the other hand, are a vast collection of slippery molecules that do not dissolve much in water but like to pack together with other lipids. A million lipid molecules form something like the crowd in the streets of Buenos Aires after Argentina won the World Cup in 2022—a mass of bodies tightly packed, but with individuals still moving through the crowd, so that who's next to whom changes from minute to minute. In the same way, lipids pack closely together to make a membrane that seals off and protects a cell or an enveloped virus.

At Syntex, Phil found that building lipid delivery vehicles for nucleic acids was really challenging. Most lipids in natural biological membranes are negatively charged, as are nucleic acids such as DNA and RNA, which means that they repel each other. So, Phil tried synthesizing positively charged lipids for packaging nucleic acids. On the one hand, he found that such lipids stuck to the nucleic acids quite nicely; on the other, they also stuck to every cellular membrane in sight—not good for making virus-like packages that could circulate through an animal. Nevertheless, he persevered and finally came up with a recipe for positively charged lipids that formed little containers called *liposomes* with nucleic acids bottled up on the inside.

Phil's progress at Syntex was derailed in 1988, when his boss explained that the lipid project was being shut down because it did not have near-term profitability for the company. Syntex ventured that the technology was more appropriate for the distant future—the year 2020. (This wild guess turned out to be prescient, because mRNA packaged in lipid particles was approved by the FDA for the Covid-19 vaccines in the third week of December 2020.) Phil and his passion for delivering nucleic acids with lipid particles then relocated south, and he founded Vical, Inc., in San Diego. It wasn't too long before Bob Malone and Inder Verma of the Salk Institute came calling.

GLOWING IN THE DARK

Malone and Verma were right that use of mRNA as a drug or vaccine would avoid one safety concern with DNA-based therapies: the possibility of making irreversible and potentially harmful changes to the genome. But they also knew that mRNA came with its own downsides, including its vulnerability to ubiquitous RNA-destroying enzymes—ribonucleases—in the body, along with the difficulty RNA had entering cells. They saw in Phil's new liposomes the potential to solve both of these problems, so the three researchers struck up a collaboration.

For their first experiments, the Salk group chose the mRNA encoding a protein called firefly luciferase, not because it had any therapeutic value, but because it was as easy to detect as firefly flashes on a dark summer's night. They mixed the mRNA-liposome agent with many different types of cells—human, mouse, rat, frog, and fruit fly—growing in Petri dishes. When the cells started glowing like minuscule fireflies, the scientists knew that the mRNA had gotten in and was being translated into protein in its new home. Thus, by 1989, they had clearly shown that mRNA could be used to reprogram cells to make a foreign protein.

The next step was to see if this worked in a living animal, not just a Petri dish. The following year, Bob Malone and Phil Felgner worked with Jon Wolff of the University of Wisconsin on experiments in which they injected luciferase mRNA into the muscle of mice. Although the mice didn't glow in the dark, the muscle cells near the site of injection did. This provided "proof of principle" that foreign mRNAs would instruct the synthesis of the corresponding protein in a mammal. But those interested in medical applications held fast to their skepticism. Sure, the treated tissue glowed—but firefly luciferase is very efficient at making a cell glow, so it wasn't clear that much protein was being made from the artificial mRNA. Most doubted that a therapeutic dose of protein could be produced from mRNA delivered in a lipid particle.

This skepticism continued even when two professional vaccine-makers in France took the next steps. In 1991, Pierre Meulien and Frédéric Martinon joined the leading vaccine company Pasteur-Mérieux. Based in the countryside near Lyon, this revered institution made vaccines from weakened versions of infectious viruses—either "vaccine strains" of a virus that had been developed to replicate poorly or viruses that were treated by heat or chemical insults until they lost most of their replicative punch. This may seem rather crude, but it often works—the incapacitated virus still displays on its surface the proteins that instruct the immune system to be on the lookout for a real viral attack. In fact, most vaccines in use today are still made by these methods.

Like Malone and Verma, Meulien and Martinon reasoned that it might be much more efficient to vaccinate with a nucleic acid and let the human body do the work of decoding the nucleic acid to make the viral protein. But which nucleic acid would make a better vaccine—DNA or RNA? Although the difficulty of producing RNA and keeping it stable was a bit daunting, Meulien and Martinon were encouraged by Malone's and Wolff's success. For them, the tiebreaker between mRNA and DNA was the safety issue—foreign DNA might become integrated into a human chromosome with unknown consequences, but mRNA did not integrate into chromosomes; the mRNA would direct the synthesis of proteins for a while and then be swept clean by cellular ribonuclease enzymes.

Meulien and Martinon aimed to use mRNA to make a flu vaccine. As an initial test, they synthesized mRNA encoding a flu virus protein, encapsulated the mRNA into liposomes, and injected the mixture under the skin of mice. They were thrilled to see that many of the injected mice developed robust killer T lymphocytes* that targeted mouse cells

---

*"T" means that these immune cells are produced in the animal's thymus, and "lymphocytes" means white blood cells.

infected by the flu virus. When such *T cells* identify a cell infected by a particular virus, they literally punch holes in the infected cell and destroy it. Importantly, killer T cells can give particularly long-lasting protection against a virus, ranging from many months to many years, depending on the virus in question. Meulien and Martinon, however, were met with lukewarm enthusiasm by the scientific community and by investors, who considered RNA to be so unstable that they continued to bet on the future of DNA vaccines.

Phil Felgner's pioneering lipid particles certainly worked for cultured cells, but when tested in animals, the lipids were found to cause several problems, including profound drops in disease-fighting white blood cells, problems with blood clotting, and severe inflammation. Scientists assumed that these adverse reactions were the result of the lipids carrying a positive charge (as positively charged lipids don't occur in

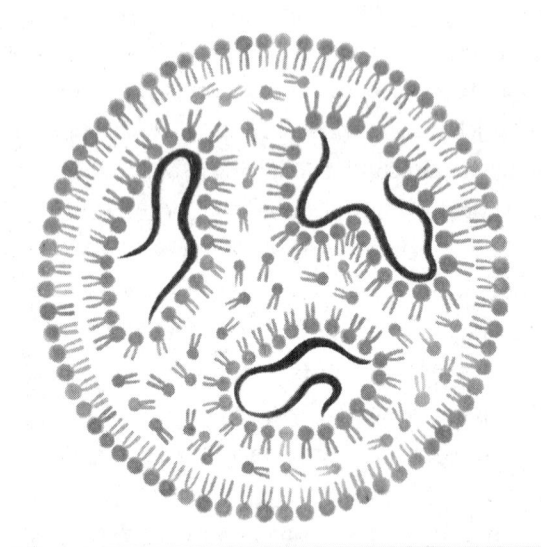

A lipid nanoparticle (LNP), similar in size to a coronavirus, is shown in cross section. Each lipid has a positively charged head group (small circle) that binds to the mRNA and two fatty "tails" that pack together with each other. Actual LNPs are made with mixtures of lipids, not the single type of lipid shown here for simplicity. Each LNP encapsulates several mRNA vaccine molecules (dark strands).

nature). A new generation of lipid formulations would be required. These would ultimately come from Pieter Cullis of the University of British Columbia in Vancouver and the companies that he founded.

First developed in 1990, these new lipids have the property of shifting their electrical charge depending on the acidity of their environment. In a slightly acidic solution, such as that produced by adding vinegar or lemon juice to a recipe, they carry a positive charge—perfect for binding to the negatively charged RNA and forming a tiny package called a *lipid nanoparticle* (LNP). When a researcher neutralizes the acidity of the solution, the electrical charge on the lipids switches off. A lack of charge helps the lipid nanoparticle circulate through the bloodstream, stick to a cell, and sneak inside. Once the package is inside the target cell, the more acidic environment flips the lipid charge back to positive. Critically, the lipids inside a cell are negatively charged. Now, negative attracts positive, the liquid nanoparticle ruptures, and its RNA cargo is released into the cell's cytoplasm.

Lipid nanoparticles developed by Cullis were used by Tom Tuschl and Phil Sharp's company, Alnylam, to deliver mRNA-cleaving siRNAs as therapeutics for hereditary ATTR in the company's first clinical trial, which came to its successful conclusion in 2018. It turned out that siRNA delivery would be a dress rehearsal for mRNA delivery a few years later.

CLOAKING THE RNA

Even with the right lipid nanoparticle delivery vehicle, an mRNA vaccine still needed to outfox a protective mechanism called *innate immunity*. As we've seen, viruses are presumably as ancient as the organisms they attack, so evolution has had a long time to develop antiviral strategies. One of these, innate immunity, is found in all animals, from worms and insects to mice and humans. It is "innate" because it doesn't depend on prior exposure to the invader. (In contrast, *adaptive immunity*, which involves antibodies and T cells, is much more specific and requires such prior exposure.) The innate immune system recognizes viral RNA because it has features that distinguish it from normal human RNAs. For example, intermediates in viral RNA replication are double stranded, formed when a plus-strand is copied into a minus-strand or vice versa. Long double-stranded RNA is rare in uninfected cells. Furthermore, viral RNAs contain plain-vanilla A, G, C, and U, whereas cellular RNAs are decorated with various nucleotide modifications—small chemical groups attached to some of their bases.

Recognition of these hallmarks of viral RNA allows the innate immune system to give us antiviral protection day after day, but it becomes a liability for mRNA vaccines. The same innate immune response that senses viral RNAs as being foreign can also sense the injected mRNA in the vaccines as being foreign, leading to a nasty inflammatory response, including rash, fever, headache, and joint pain. Not surprisingly, innate immunity has trouble telling whether incoming RNA has a beneficial intent or devious designs. Disguising mRNA so it didn't look so much like viral RNA therefore emerged as another key puzzle piece needed for mRNA vaccines, which brings us to Katalin "Kati" Karikó.

Born in Hungary in 1955, Karikó had been fascinated by biochemistry from the age of five when she watched her mother make soap from animal fat and lye. But after getting her PhD, Karikó realized how few opportunities existed for research scientists in communist Hungary. So in 1985, at the age of 30, she escaped to the United States with her husband and two-year-old daughter. The little cash they had was sewn into her daughter's teddy bear.

By 1990, Karikó had worked her way up to adjunct professor at the University of Pennsylvania. She was passionate about developing mRNA as a therapeutic. But this approach seemed far-fetched to government granting agencies. "Every night I was working: grant, grant, grant. And [the decisions] came back always: no, no, no." The goal also seemed far-fetched to her department at Penn, which demoted her to a lesser position in 1995.

Then, three years later, serendipity struck at the Xerox machine. Before the advent of electronic journal subscriptions, scientists would photocopy journal articles from the library for their evening reading. Karikó would frequently vie with a new assistant professor, Drew Weissman, for access to the copy machine. After a few tussles, they each saw what the other was photocopying and realized that they had matching interests.

Weissman was an immunologist, looking for an opportunity to improve human vaccines. Karikó was an RNA scientist, convinced that mRNA was an underappreciated shortcut to producing therapeutic proteins. Not only was their science complementary; their personalities were too. Karikó was talkative and bubbly, Weissman more reserved and methodical.

Together they devised a way of disguising mRNA so the innate immune system didn't recognize it as viral RNA. They discovered that the letter U of the RNA alphabet was the main feature that the innate immune system used to recognize RNA—presumably because it's the one letter that's unique to RNA compared to DNA. So, in 2005, Karikó and Weissman tried substituting every U in an mRNA with various modified versions of U, and they found that several of the modified versions—including one called *pseudoU*—were largely ignored by the innate immune system.

Importantly, the T7 RNA polymerase Bill Studier had pioneered for scientific use happily accepted this modified U, incorporating it at the correct places in the growing RNA chain. The ribosome still accepted it as well, reading it like a normal U. In fact, protein synthesis seemed *more* efficient with the mRNA containing pseudoU. But it got even better: like the immune system, cellular ribonucleases, too, had trouble recognizing RNA containing pseudoU, so the modified RNA was more stable than the natural version. This was almost too good to be true: the two activities that needed to be preserved—transcription and translation—were maintained or enhanced by pseudoU, whereas the two undesired activities—stimulating innate immunity and degradation—were inhibited.

Among the companies that licensed Karikó and Weissman's technology were BioNTech and Moderna. These biotechnology outfits initially aimed to use it to fight cancer. Nothing, however, changes your plans like an earth-shaking pandemic.

ASSEMBLING THE PUZZLE PIECES

On January 10, 2020, Prof. Yong-Zhen Zhang of Fudan University in Shanghai posted the RNA sequence of a new coronavirus on an open-access website. The importance of this community-minded act was not immediately appreciated, because the new virus was generating only limited concern outside China at the time. Yes, it was related to the coronaviruses responsible for two previous outbreaks of severe acute respiratory syndromes—SARS in 2002 and MERS in 2012—but each of those had been contained, resulting in fewer than 1,000 deaths worldwide. Yet the new coronavirus, soon to be named SARS-CoV-2, had a different destiny. It would be the scourge of the planet.

Somehow, the scientists at the then-obscure biotechnology company Moderna, based in Cambridge, Massachusetts, realized that same month that this new coronavirus posed more of a threat than SARS or MERS. Scientists at the equally obscure company BioNTech in Mainz, Germany, were also reading the reports of the new infections in Wuhan, China, and saw the hallmarks of an incipient pandemic: many infected but asymptomatic individuals who would unwittingly spread the virus and no travel restrictions to contain the outbreak. Both companies had been developing messenger RNA for therapeutic purposes. And both companies thought that the mRNA technology they were developing might be quickly retooled to make a protein that would serve as a vaccine against the new virus.

In many respects, the companies were making a very bold move, considering that the utility of *any* mRNA vaccine was then unproven. But they had all the puzzle pieces they needed lying on the table in front of them. For six decades, scientists had been uncovering the mysteries of mRNA. They had deciphered the genetic code, so anyone could read Yong-Zhen Zhang's SARS-CoV-2 sequence and understand how to make the Spike protein. They had shown that they could in fact use mRNA to make enough protein to elicit an immune response, central to vaccine development. They had developed a powerhouse technique for copying DNA into gobs of mRNA. They had learned that lipid-RNA combinations helped RNA enter human cells, and they'd developed the tiny greaseballs called lipid nanoparticles. And they'd discovered that the U base of mRNA could be substituted with a modified form to disguise the mRNA so it wouldn't elicit an unhealthy inflammatory response.

Yet, as any of us who have assembled a jigsaw puzzle know, having all the pieces laid out on the table is only the beginning of the hard work. To illustrate how challenging it was to produce a successful Covid-19 vaccine, consider that the mRNA vaccines were in a race with more than a dozen other contestants—many using proven technologies that seemed very likely to work once again. These multiple approaches cast a wide net: some used inactivated SARS-CoV-2 viruses, others engineered a harmless virus to express the Spike protein, still others were DNA vaccines. Some of these approaches—such as the Oxford-AstraZeneca DNA vaccine that was initially used in the United Kingdom—ended up yielding quite respectable vaccines that simply fell short of the efficacy that would be achieved by the two mRNA vaccines. Other approaches failed to elicit a strong enough immune response in humans and were dropped.

Assembling the mRNA vaccine puzzle required remarkable talent, creativity and fortitude, and some truly remarkable scientists to make it happen. Among these, the story of Ugur Sahin and Özlem Türeci is particularly compelling. Born in Turkey, Ugur Sahin moved to Germany when his father got a job at a Ford automobile factory in Cologne. Özlem Türeci was also of Turkish heritage—her biologist mother and surgeon father had immigrated from Turkey to Germany. Sahin and Türeci met in 2001 when they were both working as physicians at a hospital in the Saarland district of Germany. They married in 2002 and had a daughter. Beyond their home life, their shared passion was to bring novel science to bear on unmet medical needs, particularly in the area of immuno-oncology: stimulating the immune system to recognize and destroy tumor cells.

In 2008, Sahin and Türeci founded BioNTech with the goal of developing cancer vaccines based on mRNA (more on this later). The work was challenging, but they made progress over the next decade—they had more than a dozen compounds in clinical trials—when the fateful day in January 2020 led them on their new mission.

The Covid-19 vaccine's target would be the telltale spikes that give the coronavirus its crown-like appearance. The 90 protein spikes protruding from the fatty lipid envelope that surrounds the coronavirus are the first things the immune system encounters that warn of an incoming coronavirus attack. Stimulation of the immune system with the telltale Spike protein, therefore, should be enough to enable the immune system to recognize the real virus right away. Furthermore, because the Spike protein helps the virus enter human cells, antibodies against it—which would work by binding and covering it up—should also help inhibit viral infection.

Knowing the sequence of the new coronavirus RNA was essential to designing an mRNA that would encode this viral Spike protein, but it was only a start. For one thing, the form of the Spike protein was not constant; as the virus fused with a human cell, the spikes of the crown would flip into a different shape. If the Spike protein specified by the mRNA vaccine underwent this shapeshifting, the immune system might be trained to be on the lookout for the wrong shape. The antibodies that formed wouldn't match the coronavirus's spikes when it had just entered the body and when there was still time to prevent it from infecting us—which would render the vaccine useless. The solution was to swap into the Spike protein sequence a pair of particularly inflexible amino acids called prolines—a trick that had been developed for the related MERS virus Spike protein—thereby locking the shape in place.

Sahin and Türeci also had to decide what codons to use to encode the Spike protein. This challenge would arise with any mRNA vaccine, not just Covid-19. Because of the "redundancy" of the genetic code—most amino acids can be specified by several codons—many trillions of possible combinations would encode the same protein. Some sequences, however, would be translated more efficiently than others. Sahin and Türeci's experience working on mRNA cancer vaccines gave useful guidance, and they settled on 20 mRNA sequences to test.

But there were more questions to answer. How many doses would be required to effectively stimulate the immune system? How should their vaccine be stored, and at what temperature? How could they possibly organize human clinical trials in short order? Here, a phone call to Pfizer's head of vaccines, Kathrin Jansen, quickly brought Pfizer's enormous vaccine experience into a partnership, to the benefit of both companies—and the world.

In November 2020, the directors of Pfizer Inc. were waiting in nervous anticipation. They were about to hear the results of the clinical trial of the mRNA vaccine they were developing with BioNTech. As the vaccine's 95 percent efficacy rate was announced, a collective gasp arose—and then the group erupted with applause and shouts of triumph. The board had been hoping for at least 70 percent efficacy, which would have been a public health success. That would have put the Covid-19 vaccine somewhere between the influenza (flu) vaccine (averaging 40 percent effective, with a year-to-year range of 10 to 60 percent) and the measles vaccine (97 percent effective). Ninety-five percent was beyond most expectations. This conservative pharmaceutical company, which had a reputation for being risk-averse, had taken a bet on this unproven mRNA technology—and it had just paid off.

Similar celebrations were undoubtedly occurring in Mainz, as well as in Cambridge, Massachusetts, because the Moderna vaccine trials read out at about the same time and likewise revealed a 95 percent efficacy rate. Considering that BioNTech/Pfizer and Moderna worked independently and made many decisions differently—such as which codons to use to encode the Spike protein—it is rather amazing that they arrived at the finish line almost simultaneously and with similarly effective vaccines. It had taken 30 years for mRNA therapeutics to evolve from being generally disparaged—"too unstable," "too difficult to get into cells," "too immunogenic"—to being heralded as "A shot to save the world."

THE THERAPEUTIC OF THE FUTURE?

Over the course of the pandemic, we poured unprecedented resources into the development of mRNA vaccines for Covid-19 and generated oceans of data about their use and efficacy. We also learned about the limitations of such vaccines: how vaccinated people continue to get infected, although typically with low severity; how difficult it is for a vaccine to keep up with a virus that mutates so rapidly. Nevertheless, one could hardly have dreamed of a more impressive test case for a new drug. Indeed, the Nobel committee recognized the future potential of mRNA vaccines when it awarded Karikó and Weissman the 2023 Nobel Prize in Physiology or Medicine. Their breakthrough not only enabled the development of effective vaccines against Covid-19 but also, in the words of the Nobel Assembly at Karolinska Institute, which awarded the prize, "fundamentally changed our understanding of how mRNA interacts with our immune system," paving the way "for vaccines against other infectious diseases." Might delivering proteins by injection of their mRNA provide solutions to unmet medical needs well beyond vaccines? Are we on the cusp of an mRNA therapeutics revolution?

The truth is we don't yet know. There is certainly substantial unmet need for more, and more efficient, vaccines against other viral diseases. For example, current flu vaccines are limited in their efficacy because they can't match the diversity of the flu virus—more than 60 subtypes called strains. Vaccination against any one strain protects you from infection mostly by that strain. Each February, the World Health Organization reviews surveillance data from around the world and tries to predict which viral strains will be most prevalent during the upcoming flu season. Vaccine effectiveness drops if too many strains are mixed, so vaccines are directed against only three ("trivalent") or four ("quadrivalent") strains. Flu vaccines are mostly made by hand-injecting a vaccine strain of the virus into chicken eggs. (This is why your pharmacist wants to know if you're allergic to eggs before they vaccinate you, because small amounts of egg protein persist in the vaccine.) The six months it takes to produce the vaccine prevents it from being tailored very accurately to the strains of flu virus that will emerge next. In other words, by the time we really know which strains predominate in a given flu season, it's too late to change the vaccine. Now that we have the platform for mRNA vaccine production, thanks to the Covid-19 vaccines, there's an opportunity to whip up vaccines much more quickly than by injecting chicken eggs. The goal is to produce vaccines more tightly matched to a given year's virus, potentially leading to high efficacy. And think of the extra 140 million chicken eggs we'd save each year in the United States alone—that's a lot of omelets.

But the next big test cases for mRNA vaccines will be aimed not only at other viruses but also at cancer. Recall that both BioNTech and Moderna had been pursuing cancer vaccines before they got distracted, much to our benefit, by Covid-19.

How would such a vaccine against cancer work, even in theory? The answer may not be obvious. Viral vaccines make sense—the virus is a foreign invader, distinct from human biology, so it follows that the human immune system would recognize it as foreign and work to destroy it. The SARS-CoV-2 Spike protein is a good case in point: it is specific to the virus and is completely absent in an uninfected human, so the protein provides an unambiguous red flag warning. There are a few cancers that are known to be caused by viruses—for example, most cervical cancer is caused by human papillomavirus—so once again, it's straightforward to understand how vaccination with Merck's Gardasil is saving so many thousands of lives once lost to cervical cancer: no viral infection, no cancer. Most cancers, however, are not caused by a virus but by normal cellular processes gone awry.

A cancer vaccine is possible because tumors make aberrant proteins that are not found in healthy human tissue. Mutations in DNA—which are caused by mutagens such as cigarette smoke (lung cancer) or ultraviolet light (melanoma)—produce mutated proteins that may drive cancer. As part of their immune surveillance, cells have a natural mechanism for chopping proteins into little pieces and presenting them on their outer surface, where they can be surveyed by T cells. If the protein pieces are mutated, the T cells see them as "foreign" and they kill the presenting cell. The basic logic is that, if a cell is observed to be producing a viral protein or a mutated protein, that cell is likely infected or cancerous—and in either case, it needs to be sacrificed for the good of the organism.

In the early 1990s, the immunologist Eli Gilboa at Duke University explored whether one could ramp up this natural system and proactively train an animal's immune system to recognize and destroy tumor cells using mRNA. He conducted a series of tests on mice that had been genetically engineered to develop metastatic lung cancer. His experimental design had two novel features. First—rather than manufacture a specific mRNA—he isolated *all* the mRNA from mouse lung tumors, the idea being to train the immune system to recognize and respond to a spectrum of mutated proteins. Second, instead of injecting the mice directly with the mRNA encased in liposomes, he first isolated a special class of their immune cells, treated those with the mRNA-liposome combination, and reintroduced the cells back into the mice. The notion here was to put the mRNA right where it needed to be to train the immune system, rather than hoping that the mRNA would find the right cells within a living mouse.

Gilboa's results were impressive. The mice whose immune cells had been treated with tumor mRNA were protected when subsequently injected with tumor cells. The spread of the lung cancer was dramatically curtailed. Yet what is good for a mouse doesn't always translate directly into something that's good for humans. There have been more than 50 human clinical trials for various mRNA-based cancer vaccines, and there's no approved therapeutic yet. But now, with all the experience gained from developing the Covid-19 vaccines, companies such as Moderna and BioNTech have reenergized their mRNA cancer vaccine programs. In 2022, Moderna in collaboration with Merck announced extremely promising results for an mRNA vaccine aimed at melanoma. So success may be on the horizon.

How much of an impact are mRNA therapies likely to have outside of the vaccine space? Most traditional pharmaceuticals as well as siRNA therapies act by inhibiting a disease-promoting process, but mRNA's potential is in the opposite direction—restoring a functional protein that is absent or mutated in a patient. Diseases caused by mutated proteins include sickle cell disease, muscular dystrophy, cystic fibrosis, and spinal muscular atrophy (the last being the same disease Adrian Krainer helped treat with antisense RNA). Beyond these, the universe of so-called rare diseases, each of which afflicts sometimes a thousand or 10,000 people worldwide, provides important opportunities. mRNA therapies offer the hope that we might be able to create a single mRNA therapeutic platform into which we'd insert any of an enormous number of coding sequences, personalized to the patient, which their body's machinery would then translate into the corresponding, desperately needed protein.

Finally, consider therapeutic antibodies, which redefined the pharmaceutical industry in the 1990s. Therapeutic antibodies are laboratory-made versions of antibodies made by the B cells of our immune system, and they can be tailored to specifically bind to a protein target on the surface of a cell. In some cases, simply binding to the protein neutralizes a disease process, while in other cases the binding induces subsequent beneficial effects such as the death of a diseased cell. Commonly prescribed therapeutic antibodies include Humira for rheumatoid arthritis, Keytruda and Opdivo for cancer, Dupixent for eczema and asthma, and Stelara for psoriasis and Crohn's disease. Because these therapeutic antibodies are proteins, they can't be taken as a pill—our digestive systems are made to digest the proteins in our meals, and they don't discriminate therapeutic proteins from food. So therapeutic antibodies are usually delivered directly into the bloodstream. This requires the patient to go to a hospital or an infusion center and sit for an hour while the medicine is delivered in a slow drip through a needle in the arm. Such intravenous infusion is expensive, tedious, and often painful for the patient. Because every protein has its corresponding mRNA, it's conceivable that these antibodies could instead be delivered as mRNAs by subcutaneous injection—just like a vaccination. We don't yet know whether the mRNA route will produce a high enough dose of an antibody to be therapeutic, but the idea is attractive enough that biomedical researchers are currently exploring it. The days when investors wouldn't bet on mRNA are clearly over.

In short, the scientific principle underlying all these mRNA therapeutics is simple: in every case where a protein is needed to stimulate the immune system or to replace a missing or mutated protein, it seems possible to instead use the corresponding mRNA to instruct our bodies to make that protein. Converting the principle into practice is challenging, but the success of the Covid-19 mRNA vaccines has provided a huge stimulus.

We've now seen three types of RNAs that have been converted into effective medicines. Antisense RNA—replete with chemical modifications that improve its stability and delivery—has proved beneficial, as highlighted by its success in treating the deadly childhood disease spinal muscular atrophy. Small interfering RNA has been used to treat rare but devastating genetic diseases and may soon be deployed in the fight against neurodegenerative diseases such as Alzheimer's and ALS. And mRNA has been turned into an effective vaccine to combat the worst pandemic in living memory and is poised to provide humankind with new vaccines and therapeutics.

In all three cases, the therapy works at the level of RNA, making no change to human genes. In fact, as we saw, one of the perceived benefits of mRNA vaccines is that there is no risk of altering the genome. But now, another natural RNA-based process—a defense system that bacteria use to protect themselves against viral infection—has been transformed into a tool to edit the genome of any species, including humans, with never-before-imagined speed and specificity. Unlike earlier RNA-based interventions, this technique produces a permanent alteration in the genome, which is why it's so powerful and—if misused—potentially perilous. And, once again, it is the miracle molecule RNA that is catalyzing the revolution.

Chapter 11

RUNNING WITH SCISSORS

Imagine a world in which scientists can create heat- and drought-resistant crops that would thrive on our warming planet, or organisms to sequester carbon and reverse the effects of climate change, or quick and safe fixes for devastating genetic diseases. This is the utopian vision of a future transformed by CRISPR—the gene-editing technology recent books have touted as offering "the unthinkable power to control evolution," the awesome prospect of "editing humanity," the chance to direct "the future of the human race." Yet not everyone is so bullish on CRISPR. There are also dystopian visions of how it might change the world, in which irresponsible individuals or governments use it to create packs of attack animals or legions of obedient superhumans. Is it conceivable that someone could genetically engineer an army of Clone Troopers, as in a *Star Wars* film?

CRISPR enables us to modify the genes of practically any organism, from mosquitos to corn to human beings. While many people associate CRISPR with gene-engineering technology, its working parts were actually discovered as a natural process in bacteria, which use it to stave off attacks by the viruses known as phage that we've encountered again and again. The war between bacteria and their viruses has presumably been waged for more than a billion years, and whenever either side invents a new attack, the other comes up with a new counterattack. So, in a sense, the CRISPR system—which works against DNA-based viruses—isn't special but is only one among many antiphage protection systems.* But it's the first to have been retooled into a genetic-engineering kit, and it derives its unprecedented power from RNA.

CRISPR's DNA-cutting machinery is made up of two parts. The first is the Cas9 (CRISPR-associated 9) protein. This protein is an enzyme that works a bit like molecular scissors, physically cutting both strands of the DNA double helix. We've all been told by our parents not to run with scissors, and the same danger applies here: if unguided, these DNA-cutting scissors could do a lot of random damage. That's where RNA comes in. In a 2012 breakthrough that would win them the 2020 Nobel Prize in Chemistry, my former postdoc Jennifer Doudna, whom we last saw unlocking the secrets of RNA's structure, and her collaborator Emmanuelle Charpentier discovered that by pairing the Cas9 enzyme with a customized "guide RNA," they could precisely direct CRISPR's scissors to cut any genetic sequence. Within months after Doudna and Charpentier's discovery was announced, several research teams showed that this RNA-guided genome editing worked in living human cells; these included the groups of Feng Zhang at MIT, George Church at Harvard, and Jennifer's own lab at Berkeley. The stage was set to think about using CRISPR to inactivate cancer-causing *oncogenes* or the genes that encode misfolded proteins in Alzheimer's disease or amyotrophic lateral sclerosis, among a plethora of possible applications.

How, exactly, does RNA make such things possible? As we've seen time and again, RNA is an ace at base-pairing, doing the kind of molecular matchmaking that gets nucleic acids talking to each other in pro-

---

*The CRISPR-Cas9 system, the first to be converted into a genome-editing tool, protects bacteria against DNA phages, but there are other RNA-guided CRISPR systems that cleave RNA and protect bacteria against RNA phages.

ductive ways. The RNA molecule associated with the Cas9 protein uses this power to base-pair with one of the two strands of DNA and thereby specify with dazzling precision the exact site of genome editing. Before the advent of this technology, methods for editing specific sites within the human genome were expensive, tedious, and so difficult that only a few labs in the world were able to use them. But nowadays, with a bit of training, undergraduate students can assemble a CRISPR kit to erase a bad mutation in a gene or knock out a gene altogether in a few weeks. In my lab, CRISPR is mentioned so often that it's become a verb, like "google." Once we've "CRISPRed" an alteration into a gene, we can see what effect it has on the cells growing in our Petri dishes.

My lab is hardly alone in embracing CRISPR. While the more radical applications of this technology are rightly being debated, an estimated 7,000 research laboratories worldwide have already used CRISPR, some making impressive new discoveries. Many more breakthroughs are on the way. Researchers routinely use it to produce genetic alterations not just in cells but in living animals, such as fruit flies and mice. Scientists are generally hype-averse, but many in the life sciences describe the advent of this technology a bit like the Second Coming. They commonly call it the "CRISPR Revolution."

BACTERIA TO THE RESCUE

To understand scientists' enthusiasm and to truly appreciate the lofty promise of CRISPR, we must start with its lowly origins. Researchers first identified what we now call CRISPR simply by staring at the DNA sequences of bacterial genomes. The researchers were accustomed to identifying bacterial protein-coding genes—stretches of triplet codons, each beginning with an ATG start codon and ending with one of the three stop codons. But the CRISPR DNA caught their attention as something new and unusual. It contained repeated sequences that were palindromic, reading the same forward and backward, like "Madam I'm

Adam." Interspaced with these repeats were snippets of sequence that, it emerged, came not from bacteria but rather from phages, the viruses that attack them.

It seemed that the bacteria were keeping track of previous encounters with invading phages. The repeats were later found to direct the insertion of new phage sequences into the array, akin to updating one's list of contacts. These DNA arrays were given a very long name—clustered regularly interspaced short palindromic repeats—which researchers mercifully abbreviated as CRISPR. But the purpose they served remained unclear.

Jennifer Doudna had been introduced to CRISPR in 2006 by her Berkeley colleague Jill Banfield. Banfield was initially intrigued by these unusual DNA sequence arrays, but the emerging functional implications were even more exciting. Scientists in the yogurt industry—which relies on healthy bacteria to convert milk into yogurt—found that CRISPR allowed their probiotic bacteria to obliterate attacking phages. Soon after, they and others found that the CRISPR system cleaved the incoming phage DNA to destroy it. Once a bacterium had a snippet of a particular phage sequence stored in its CRISPR cluster, it was immune to attack by a phage of that type. It was as if CRISPR allowed a bacterium to shout, "My ancestors have seen you before and you are really bad, so now I'm going to destroy you." CRISPR was rather like the human immune system, which is primed by past infections or vaccinations. But CRISPR was even better, because the bacteria had the immunity embedded in their DNA so it was passed from one generation to the next. The bacteria were essentially born prevaccinated.

Now RNA entered the scene. A Dutch research group headed by John van der Oost found that small RNAs were transcribed from the CRISPR repeats in bacteria, and these somehow directed the antiviral defense. But how? With her expertise being RNA, and RNA apparently playing this key role in CRISPR, Jennifer decided to direct a portion of her lab's efforts to understanding how it worked. It didn't take her long to become a leader in this nascent field.

By 2011, Jennifer suspected that the CRISPR RNAs were the key to the specificity of this antiviral system. They seemed to be acting as guides that could identify which phage DNA sequence needed to be cut while leaving the bacterial chromosome—which didn't have a matching sequence—untouched. But if RNA had the job of identifying the DNA to be cut, what was responsible for the actual cutting? The answer to this question would arise from a collaboration between Jennifer and Emmanuelle Charpentier from Umeå University in Sweden.

Emmanuelle is a slight woman, but when she steps up to a podium, she is so articulate, focused, and clear-thinking that any audience is immediately riveted. Jennifer had admired Emmanuelle's work from afar, but it would take a fortuitous conference in Puerto Rico in 2011 to bring them together for the first time. When they met, Jennifer found Emmanuelle to be soft-spoken, slyly humorous, and refreshingly lighthearted. Enjoying each other's company often portends well for a collaboration.

Emmanuelle had found that if one mutated a protein called Cas9 in the bacterium that causes strep throat, the bacterium was no longer protected from phage attack—CRISPR no longer worked. Could this Cas9 protein, then, be the hitherto unknown enzyme—the scissors—responsible for cutting up invading phage DNA? And, if so, how did RNAs copied from the CRISPR spacers show those scissors where to cut? These questions were mutually exciting to Emmanuelle and Jennifer, so they agreed to tackle them together.

To test whether the Cas9 protein was indeed the scissors for CRISPR, they decided to purify the protein and mix it in test tubes with various potential guide RNAs. They would then add various synthetic DNAs. Some of these DNAs would be copies of phage sequences matching a guide RNA (the targets), and others would be nonmatching DNAs (the controls). If their hypothesis was correct, the phage-like DNA would be cut in half in the presence of the matching guide RNA, but nonmatching DNAs would be left intact.

In Jennifer's lab, the research was coordinated by Martin Jinek, a modest and superbly talented Czech postdoctoral fellow. Martin's counterpart in Emmanuelle's lab would be a Polish graduate student named Krzysztof Chylinski. Fortuitously, both had grown up near the Czech-Polish border and spoke Polish, facilitating their frequent Skype conversations.

The first step was to purify the protein. Krzysztof mailed the gene for Cas9 to Martin, who immediately set to work trying to get *E. coli* to make the protein. Expressing a protein from a different bacterium in *E. coli* often worked but was not guaranteed, so Martin and a student working with him had to explore multiple conditions before they optimized Cas9 protein production. Then the key experiment: mix the Cas9 protein and the guide RNA together in a test tube so they could bind each other, add DNA with a sequence that matched that of the guide RNA as a stand-in for the phage DNA, and look to see whether that DNA, and only that DNA, was cleaved.

And the experiment . . . totally failed. The target DNA emerged from the reaction unscathed.

The perplexed collaborators held a brainstorming meeting by Skype. Perhaps the hypothesis was wrong, and Cas9 was not the DNA-cleaving enzyme after all? Or could there simply be some ingredient missing in the molecular recipe? Emmanuelle had found a second RNA, dubbed *tracrRNA* for "trans-activating CRISPR RNA," that was required for the strep bacteria to produce CRISPR guide RNAs. Might this second RNA also be required for Cas9 to cleave DNA?

Martin tried a new test-tube mixture with both the guide RNA and the tracrRNA. This time the target was cleanly cut into two pieces. In order for the Cas9 and guide RNA to function like a pair of genetic scissors, they needed help from the tracrRNA; it played the role of a thumb, holding the scissors in place to make the cut.

Not only was the target DNA cleaved but also the specificity was exquisite. DNA molecules that didn't match the sequence of the guide were untouched, but the uncut DNA could be cut simply by designing a new guide RNA to match 20 nucleotides of its sequence. It was similar to the "Search" function in word-processing software: if you search

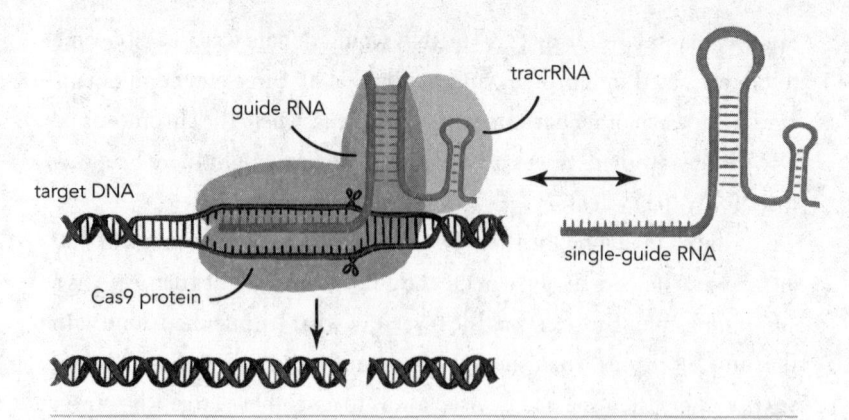

The CRISPR-Cas9 bacterial defense complex (left) has a guide RNA that recognizes a target phage DNA sequence and a tracrRNA that base-pairs with the guide RNA and holds on to the Cas9 protein. Cas9 then cleaves both strands of the DNA. An engineered form of CRISPR-Cas9 used for genome editing (right) has the two natural RNAs fused to form a single-guide RNA.

for a string of 20 letters, the software will highlight any text that's a perfect match.*

Jennifer and Martin were not the sort of scientists who rest on their success. So when Jennifer sat down with Martin to look at his new data, she congratulated him on his discovery, then turned the conversation to: "How can we build on this?" Requiring *two* separate RNAs and a protein for CRISPR cleavage clearly worked just fine in the test tube and in bacteria. But looking ahead to how the technology might eventually be used in human cells, it seemed complicated. You would have to introduce *three* pieces of DNA into the cell, relying on the cell to manufacture three components: the Cas9 protein, the guide RNA, and tracrRNA. Then you would have to basically hold your breath, hoping that all three pieces of DNA would get transcribed into their respective RNAs by an RNA poly-

---

*Subsequent work showed that the CRISPR system did not require a perfect 20-out-of-20 match between the single-guide RNA and the DNA target. This leads to some "off-target" editing, a potential concern that scientists have been working to address.

merase, that the resulting Cas9 mRNA would find its way to a ribosome to be translated into protein, and finally that all three end components would find each other back in the cell's nucleus, where the chromosomal DNA target awaited. It was a lot of moving parts. Might there be a way to simplify the system?

It didn't take long for these two experienced RNA scientists to find an answer. They saw a way to make the guide RNA and the tracrRNA as a single molecule. The trick was the reverse of what Jennifer had done with the SunY ribozyme when she was a grad student in Jack Szostak's lab 25 years earlier. Then she had devised a way to assemble a large RNA from several pieces of more tractable size. Now, looking at the two CRISPR RNAs, Jennifer and Martin saw how they could link them together to make a *single-guide RNA*. When this dynamic duo—Cas9 protein plus single-guide RNA—cleanly cut its DNA target in the test tube, the Doudna-Charpentier team was ready to publish their discovery—and brace themselves for the tsunami of interest that would immediately engulf them.

FIRST THE SCISSORS, THEN THE GLUE

If DNA cleavage had been the beginning and end of CRISPR's magic, the interest among academic labs, much less industry, would have been much more modest. The appearance of a new tool to cleave DNA would not by itself have spawned a cottage industry of biotech companies. But the discovery—that RNA-guided Cas9 cleaved DNA with such previously unimagined specificity—did not occur in a vacuum.

Starting with research in yeast and other fungi and then moving on to fruit flies and mammalian systems, scientists already knew that living cells cannot afford to let broken DNA molecules stay broken for long. The integrity of the genome is essential for life. A long, long time ago, organisms found ways to patch broken DNA quickly. As a result, if an experimenter uses CRISPR to slice a gene, the cell's repair machines go to work. That's when the gene editing happens.

DNA repair in eukaryotic organisms, including humans, comes in two main flavors. The first is a quick-and-dirty "emergency repair" process that rejoins broken chromosome ends. Its technical name is *non-homologous end joining*, or NHEJ. The "end joining" part is self-explanatory. "Non-homologous" means that the two ends don't need to have any DNA sequences in common—any two broken DNA ends can be rejoined. The hallmark of NHEJ is that the stitching-together process is sloppy, resulting in a few nucleotides being lost or inserted at the site of repair. In other words, this process often leaves the repaired DNA with some unpleasant mutations. Thus, CRISPR cleavage followed by NHEJ often inactivates a gene by messing up the orderly string of codons that specify a protein.

Our trusty word-processing analogy can help demonstrate how this works. The single-guide RNA locates a DNA sequence like the "Search" function locates a string of letters in a document. We type in a sequence of letters—*The big cat*, encoded by ATGCCTTCG—and the software finds an exact match:

```
GTAGGGC ATG CCT TCG AAA ATA TTT TGT TAG CGC CTC CTT GGA GTA GAA
        The big cat ate one fat rat.
```

Once again, the start and stop codons are italicized. Now that the guide RNA has located the site of action, the Cas9 enzyme cleaves the DNA at that spot, akin to hitting "Return" on our keyboard. The text is interrupted by a line break:

```
GTAGGGC ATG CCT TCG
                    AAA ATA TTT TGT TAG CGC CTC CTT GGA GTA GAA
        The big cat
                    ate one fat rat.
```

If we now hit "Backspace," the original text is restored. This is akin to repair of DNA by NHEJ. But recall that NHEJ is sloppy—it often inserts or deletes a letter or two by mistake before it rejoins the ends. Accordingly, we'll insert a typo—a single letter T (underlined below) before stitching the sentence back together:

```
GTAGGGC ATG CCT TCG TAA AAT ATT TTG TTA GCG CCT CCT TGG AGT AGA
The     big cat run now see fox run out big big fun sun ate
```

The insertion of even this single letter destroys the meaning of the sentence, or of the gene. Such is the price of quick-and-dirty repair by NHEJ.

The second type of DNA repair is *homologous recombination.* The broken end of the DNA essentially looks around for a matching DNA sequence—this is the "homologous" part—and then undergoes "recombination" with that sequence, using it as a template to repair itself precisely. The matching DNA sequence can come from the other copy of the chromosome; that is, the chromosome inherited from Mom can be used as a template to repair a broken chromosome from Dad. Because homologous recombination has more demanding requirements than NHEJ, it occurs less frequently.

At first glance, it doesn't seem like such perfect repair would be useful for gene editing—and it wouldn't if it simply restored the original DNA sequence. But long before CRISPR, scientists had found that they could easily trick homologous recombination by introducing* a *donor template* DNA that matched the sequences near where a break had occurred but included a patch of nonmatching DNA farther away. The nonmatching DNA would come along for the ride. That way, scientists could precisely edit the sequence of a gene. Or, even more dramatically, they could introduce a sequence that started out matching that of a gene, but then inserted some new genetic element, redesigning the

---

*Multiple methods have been developed to introduce foreign DNA into living cells. Some of them involve damaging the cell surface in the presence of the foreign DNA, so that when the cell fixes the damage, some of the DNA gets inside. Viruses can also be engineered to carry nonviral DNA into cells. In another technique, called biolistics, the foreign DNA is coated onto small particles that are literally shot into the cell.

target DNA

CRISPR cleavage

NHEJ

donor template

homologous recombination

small insertion or deletion at the repair site

new sequence inserted

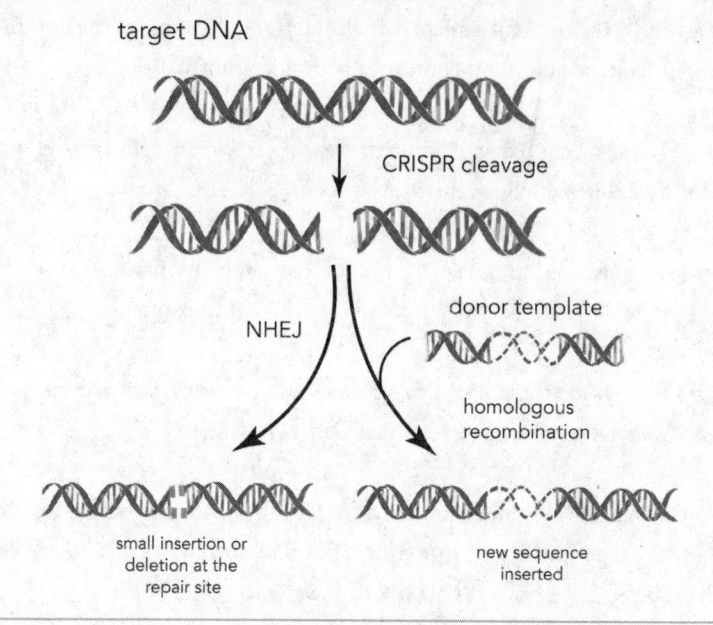

After CRISPR-Cas9 cleaves its target DNA at a specific sequence, cellular DNA-repair machines take over and do the gene editing. Non-homologous end joining (NHEJ) is a quick-and-sloppy repair system that usually leaves a small insertion or deletion of DNA at the site of repair. Homologous recombination is accurate and requires a donor template DNA to guide the repair. The donor template can introduce new sequences (dashed portion of double helix) to correct a mutation in human DNA or add a new genetic element, as long as the flanking sequences match those of the target DNA.

gene entirely. Before CRISPR, there hadn't been an easy way to make a specific break in the DNA to designate the site of recombination. But now, with RNA-directed CRISPR, it was possible to do so with spectacular accuracy.

To imagine homologous recombination in action, consider editing a document with "Copy and Paste," adding new letters, and then stitching the sentence back together. Here's the information in the chromosome:

```
GTAGGGC ATG CCT TCG AAA ATA TTT TGT TAG CGC CTC CTT GGA GTA GAA
        The big cat ate one fat rat.
```

And here's the donor template, which starts out the same as the chromosomal sequence but then continues with new information:

```
ATG CCT TCG CTT ATG TTG TTA GTA TGG TAG CGC CTC CTT GGA
The big cat and the fox run for fun.
```

CRISPR cleavage followed by homologous recombination causes the altered sentence to be incorporated into the chromosome:

```
GUAGGGC ATG CCT TCG CTT ATG TTG TTA GTA TGG TAG CGC CTC CTT GGA
        The big cat and the fox run for fun.
```

So the big cat has gone from eating a fat rat to running with a fox. Just as the "Copy-and-Paste" function allows us to rewrite a sentence in a document, RNA powers CRISPR to rewrite the code of life.

DEAD CAS9

It didn't take long for the buzz about the new CRISPR discovery to cross the Bay Bridge into San Francisco. Even before Jennifer and Emmanuelle's paper about their CRISPR-Cas9 work was published in 2012, Jennifer started brainstorming with colleagues at the University of California, San Francisco, about ways to use CRISPR-Cas9's enormous precision to switch specific human genes on or off—without cutting the DNA at all.

Why would a no-cut CRISPR system be a useful addition to a genetic toolkit? Although the guide RNA could program Cas9 to make a double-stranded DNA break at an exact site, after that the researcher was still counting on existing cellular machinery to patch the break. The repair was somewhat random—occurring most frequently by sloppy NHEJ and only occasionally by precise homologous recombination. As we've seen, repair by NHEJ has uncontrollable and sometimes deleterious consequences, and the scientists wanted to avoid that. If there's no cut, there are no DNA ends, and there's no NHEJ.

Several groups of scientists made mutations in the Cas9 protein that incapacitated its DNA-cutting activity but did not interfere with its ability to bind a single-guide RNA. They dubbed their creation "Dead Cas9" for short. They now found that by coupling a variety of other proteins to Dead Cas9, they could deliver those proteins to specific sites in the human genome. Among these were proteins already known to possess the ability to activate and repress genes.

The analogy is far-fetched, but imagine that you wanted to use a laser-guided missile system to send someone flowers. You would first deactivate the warhead so the missile could no longer explode (that's the "Dead Cas9" part). You could then use the missile's guidance system (guide RNA) to deliver your bouquet to their doorstep with absolute precision. You would key in their coordinates and, instead of blowing up their house, you would make their day. The point is that once you've developed a precision-guided delivery system, you're not restricted to delivering one kind of payload.

The speed with which new Dead Cas9 inventions were conceived and proved to work was astounding. Jennifer and Emmanuelle's paper announcing the original CRISPR-Cas9 system was published in *Science* on June 28, 2012, and already by December of the same year, the collaborative Berkeley-UCSF group had hooked up known gene repressors and activators to Dead Cas9 and shown that several human genes could be turned on or off on command. Other research groups quickly jumped in to engineer other Dead Cas9 tools. Research rarely moves at this pace. The talent and hard work of the scientists must of course be credited, but the previously unimaginable robustness of the RNA-guided Cas9 machine was a key component as well. When you've got a lightning bolt in your hand, it's time to strike!

David Liu, a chemist and biochemist at Harvard, used Dead Cas9 to pioneer a particularly useful gene-editing technique. He was interested in how one might correct a single-base mutation in a gene, as occurs in sickle cell disease and in many other human genetic diseases. Using first-generation CRISPR techniques, one would need to add a donor template DNA with the correct sequence and hope that homologous recombination won out over NHEJ, making the whole process possible but inefficient. David knew that scientists had many years earlier found "base-editing enzymes" that could change one of the letters of the DNA alphabet to another. He reasoned that he could couple one of these base-editing proteins to Dead Cas9 and thereby erase a single-base mutation at a site directed by the guide RNA. This would give him control over the gene-editing process instead of relying on the cell's own homologous recombination machinery to fix a gene.

David's reasoning was borne out. He found that the guide RNA could indeed direct the base-editing enzyme to any site in the genome that needed to be "fixed," and the enzyme would change a single letter—for example, a C to a T—in the DNA without ever causing a double-strand break. The new letter T would then be transcribed into a U in the mRNA (underlined in the diagram below), changing the triplet codon so that it specified the correct amino acid.

```
GUAGGGC AUG CCU UCG AAA AUA UUU UGU UAG CGC CUC CUU GGA GUA GAA
        The big cat ate one fat rat.
GUAGGGC AUG CCU UUG AAA AUA UUU UGU UAG CGC CUC CUU GGA GUA GAA
        The big fox ate one fat rat.
```

While just one letter may seem like a small change, in some cases it can mean the difference between life and death.

This is a thrilling time for CRISPR research, as new innovations are being created every year. The original gene-editing technique—in which Cas9–guide RNA is used to cleave DNA near a target site, and DNA is supplied as a donor template for homologous recombination—is still very much in play and is being improved all the time. Various Dead Cas9 strategies, taking advantage of the "homing" power of Cas9–guide RNA but avoiding DNA cutting, are also moving forward; David Liu's base editor is one of several approaches. Other scientists are investigating alternative CRISPR systems that use relatives of Cas9—including a protein enzyme called Cas12a—that appear to cut DNA in a way that skews the odds of DNA repair in favor of homologous recombination over NHEJ. This puts us in the fortunate position of having a diverse CRISPR toolkit, meaning that if one of the approaches should fall short as a therapeutic, we have backup plans.

CRISPR THERAPIES

Our genetic scissors allow us to make precise edits to the genome—but *what* should we edit?

The first, most obvious answer is a mutation that causes a genetic disease. As we've seen throughout this book, numerous human genetic diseases are caused by a localized mutation in a gene. The first example pinned down at the molecular level was sickle cell disease, where a single base-pair mutation in the beta-globin gene—encoding a subunit of the blood protein hemoglobin—leads to the substitution of an amino acid called valine for one known as glutamate. Everything else about the hemoglobin proteins is normal, so the mutant proteins work pretty well most of the time, carrying oxygen around the bloodstream. But suddenly—triggered by stress, or dehydration, or exercise, or infection—the mutant proteins will clump together inside the red blood cell, causing the cell to distort from its normal saucer shape to an elongated sickle shape. These distorted blood cells stick to each other, blocking blood vessels and causing a crisis that can be so painful as to require hospitalization. The hospital can try to manage the pain or perhaps even give the patient a blood transfusion, but there's no cure.

Other human genetic diseases caused by localized mutations include beta-thalassemia (as we saw, the first disease recognized to be caused by mis-splicing), Tay-Sachs disease, cystic fibrosis, and many forms of muscular dystrophy. Like sickle cell disease, these are all incurable diseases. Therapies based on RNA that we discussed in earlier chapters—such as siRNA and mRNA—might serve as useful treatments for these illnesses, but they cannot fully cure them because they wouldn't completely erase the mutant protein. By editing the gene back to its normal, unmutated state, one could in theory find a cure. Using CRISPR, such gene editing has already been accomplished in fruit flies and in mice. Why not in humans?

In considering CRISPR gene-editing as a therapy, though, there is much for biomedical scientists to consider. For one thing, where should they start? After all, it takes years of effort by teams of researchers and an investment approaching $1 billion to develop a therapeutic for a disease, even with the precision and time-saving advantages offered by CRISPR, so one needs to choose a target disease carefully. One important factor would be the lack of an effective treatment for the disease, meaning that successful gene editing would make a huge impact.

Next, the candidate disease should be really debilitating or even deadly, so that the benefit-to-risk ratio is as high as possible. Furthermore, a good candidate would be one where partial restoration of the correct protein would be therapeutic, as it's not yet possible to edit the DNA in all the affected cells in the body. This means the cells that are affected by the mutation should be readily accessible, which makes blood diseases particularly attractive as a target. You can pump blood out of a person's body, treat it, and pump it back in. Compare this, for example, with a disease such as Alzheimer's that affects the brain. Needless to say, getting access to the 100 billion cells inside a human brain is an incredible challenge.

Sickle cell disease meets all these criteria. It's extremely debilitating, and there's currently no cure. We know exactly which DNA base pair we would need to change to fix the hemoglobin protein. And human blood is a readily accessible tissue. So it's not surprising that most of the major CRISPR biotech companies have a sickle cell program. Each company is taking its own approach, which is good for the science—multiple "shots on goal" increase the chances of hitting the net. Although David Liu's base-editing technology using Dead Cas9 seems custom-designed for just this sort of application, it has some healthy competition.

Yet major challenges must be overcome even just to develop a CRISPR cure for sickle cell. Our red blood cells are essentially little bags stuffed with hemoglobin; they have lost their DNA, so they have no beta-globin gene to edit. These red cells are derived from blood stem cells in our bone marrow. Those cells still have all their genes—including the mutated sickle cell gene that needs to be fixed. The good news is that stem cells keep dividing to produce the cells that become red blood cells, so if we can repair a gene in a stem cell, the benefits of the repaired protein are passed on to all its daughter cells. The challenge is that these stem cells are rare and therefore difficult to isolate, and once they're collected from the patient, they're difficult to grow in the laboratory, where the gene-editing process needs to happen. The edited stem cells then need to be transplanted back into the patient so they can find a home in the bone marrow (the technical term for this is "engraftment"). Because they are the patient's own cells, there should be no immune rejection as can otherwise occur with a bone marrow donor. And, unlike with typical bone marrow transplants, irradiation may not be required.

In 2020, Victoria Gray, a mother of four from Mississippi in her thirties, became the first sickle cell patient to be treated with CRISPR. Gray had always lived in fear of sudden attacks of horrible pain. Her fatigue was often so debilitating that she couldn't care for her own children, and she would spend nights in the emergency room getting blood transfusions that offered only temporary relief. Now after CRISPR therapy, she is for the first time enjoying a healthy life. While such stories are certainly encouraging, clinical trials are needed to rigorously assess the benefits and safety of CRISPR therapeutics, and such trials are now ongoing. In 2023, a CRISPR-based drug called Exa-cel (or Casgevy) made history when it was authorized to treat sickle cell and transfusion-dependent beta thalassemia, first by regulators in Britain and then by the U.S. FDA. The hope is that this promising drug will soon be joined by other clinically tested therapeutics.

PUBLIC VALUES

If you're in possession of these all-powerful genetic scissors, you might ask why you'd restrict your efforts to correct mutations in *somatic cells*, or body cells, and not apply them to embryonic cells. Wouldn't a more efficient way to cure a genetic disease involve correcting the error before a person was even born? He Jiankui thought so. A professor at the Southern University of Science and Technology in Shenzhen, China, he became a pariah in the science world in 2018 after announcing that he had used CRISPR to genetically edit the embryos of two twin sisters, eliminating the human protein that HIV uses to enter cells in the hopes of conferring immunity against HIV infection. He was fired from his university and given a three-year prison sentence for "illegal medical practices." While he has remained defiant in the face of public outcry, in 2023 he told the British newspaper *The Guardian* that he felt he had moved "too quickly."

He's actions violated a consensus in the scientific community that we should put some boundaries around what's allowable when it comes to CRISPR and other types of genetic editing. These are "at least for now" sorts of boundaries, in that they can and should be reevaluated as data about safety and efficacy accumulate. One boundary is that human gene editing should be restricted to somatic cells and not be used in embryos or *germline cells* that give rise to sperm and egg. Genetic changes in somatic cells cannot be passed down to the next generation, whereas those made in germline cells can be inherited. The thinking here is that if gene editing should happen in the wrong place—so-called off-target editing—it should be confined to the patient who's being treated and not become a burden to future generations.

A second boundary that has wide acceptance concerns "enhancements." Examples of enhancements would include using CRISPR gene-editing to make your offspring taller, stronger, able to run faster and jump higher, or more beautiful in your eyes. Such uses could turn CRISPR into a dangerous tool for eugenics. While we can argue over the ethicality of such enhancements, most people would agree that the first priority for CRISPR resources should be treatment of severe disease. Even in this category, there are valid questions about safety and efficacy that need to be answered. And even if we agree to avoid germline editing and enhancements, and we create a policy consensus about the safety issues, some still question whether we have the right to alter the hand that Mother Nature has dealt us.

It's important to note that CRISPR therapies are the newest wrinkle in the established field of gene therapy. Five gene therapies, none of which use CRISPR, have been approved for use in the United States as of January 2023. All aim at introducing a functional copy of a gene to compensate for a mutant gene, but they are unable to control where the replacement gene will land in the human genome. For example, hemophilia B is a rare bleeding disease caused by low levels of a blood-clotting protein, and use of a virus to introduce healthy copies of the gene coding for the clotting factor has shown efficacy in clinical trials. CSL Behring, the company that developed the treatment, will charge $3.5 million for the one-time treatment, which sets a new record for the most expensive drug available. The hope is that the improved efficiency and specificity of RNA-guided CRISPR will make future gene therapies simultaneously safer and cheaper.

Aware of both the lifesaving potential and the possible pitfalls of genomic engineering, a group of U.S. biomedical leaders, lawyers, and ethicists met in Napa, California, on January 24, 2015, for an open discourse on a prudent path forward. This distinguished group reached a consensus to strongly discourage germline genome editing in the near term, leaving the door open if guidelines for responsible use of such approaches might be developed in the future. They also stressed the importance of support for open-access research to evaluate the efficacy and specificity of CRISPR in human and nonhuman applications and advocated for open forums to educate the public on the risks and rewards of CRISPR. The overarching goal of these leaders, shared by many others in the scientific community, is to prevent the loss of public trust in CRISPR technology that might occur if fear and confusion about potential risks overshadowed the rewards that could be reaped if the technology were used in a responsible and regulated way.

SWATTING MOSQUITOS—BIG TIME

The use of CRISPR to treat human patients isn't the only application that presents such a mix of big opportunities and big challenges. The same ambivalence accompanies CRISPR's many potential environmental applications. One proposal that has generated particularly heightened levels of both excitement and concern pertains to a particularly nasty mosquito.

Malaria is an enormous public health issue, killing more than half a million children each year, mostly in Africa and Southeast Asia. We know the cause: a microscopic parasite called *Plasmodium*, which requires a specific species of mosquito called *Anopheles* to carry out its life cycle. The *Anopheles* mosquito picks up the *Plasmodium* when the mosquito bites an infected human. The *Plasmodium* then reproduces within the mosquito. The next time the mosquito bites a human, the *Plasmodium* comes along for the ride, perpetuating the infectious cycle.

A variety of public health measures can help to combat malaria. *Anopheles* come out at night, after people have gone to sleep, so mosquito netting over beds offers substantial protection. Draining of swamps and other open sources of water can help by reducing the mosquito population. But a much more radical approach has been suggested: eradicate the *Anopheles* species. In principle, our RNA-guided CRISPR machinery has the potential to do just that.

The method is called "CRISPR gene drive." It involves engineering a Cas9–single-guide RNA combination that homes in on a mosquito gene that's essential for female fertility. The gene is present in both males and females, but it's only expressed in females. Identifying such genes isn't difficult, because mosquitos are somewhat related to fruit flies, and years of fruit fly research have identified female fertility genes that have counterparts in mosquitos. So when a piece of DNA encoding both the Cas9 protein and a single-guide RNA tuned to a fertility gene is injected into an *Anopheles* mosquito, the CRISPR machinery assembles itself within the mosquito and cleaves the fertility gene accordingly. So far, pretty simple—but now for the really creative part. The injected DNA is designed also to serve as a donor template for homologous recombination, such that the sequences encoding Cas9 and the single-guide RNA become embedded in the mosquito's fertility gene. Thus, this single insertion event accomplishes two feats simultaneously: it inactivates the female fertility gene, and it embeds the machinery necessary to propagate that gene inactivation into the mosquito's genetic makeup.

The result is that any female mosquito that mates with a male carrying the gene drive becomes infertile, while males remain able to breed and pass on the CRISPR machinery now embedded in their DNA. CRISPR "drives" through the population, until eventually enough females are infertile that the population dies out.

There's enough research on CRISPR gene drive to give us some confidence that it would actually work in nature. Of course, mutant mosquitos resistant to CRISPR would probably arise, given the enormous fitness advantage that any CRISPR-resistant mosquito would enjoy. Nonetheless, if CRISPR-engineered mosquitos were released into the environment, they would likely decimate the *Anopheles* population, potentially preventing a gruesome death by malaria for hundreds of thousands of children each year.

The rewards of such a potential intervention are obvious. But what of the risks? These are less clear. Consider the question recently posed by an opinion article in the prestigious *Proceedings of the National Academy of Sciences of the USA*: "Is CRISPR-based gene drive a biocontrol silver bullet or a global conservation threat?" The hypothetical environmental effects of releasing a gene drive system into the environment are not easy to assess. At first glance, one might worry that dragonflies, birds, bats, and spiders that eat adult mosquitos or fish that feast on mosquito larvae might suffer if the *Anopheles* mosquito population were eradicated. However, ecologists who have studied the food value of *Anopheles* find this scenario to be unlikely, as these predators readily eat other species of mosquitos and other insects.

More difficult to predict is the possibility that, as it's hopping from mosquito to mosquito, the CRISPR system might on a rare occasion miss its gene target, inserting itself somewhere else in the mosquito's genome. In this case, the progeny would no longer be infertile, which would make the mosquito's eradication less effective. And it's possible that such an error might somehow spread a gene that actually *increased* the fitness of the mosquito, making a terrible problem even worse.

Such nightmare scenarios might be easily dismissed as close to impossible were it not for humanity's previous dismal track record of introducing nonnative species into the environment. A government scientist bred and released cane toads in Australia in 1935 in an attempt to control sugarcane beetle infestations, and the original 102 imported toads have multiplied to a population of over a billion. These enormous toads are poisonous, and any native animals that eat them are likely to die. The cane sugar industry imported the Asian mongoose into Hawaii to control rats, but the mongoose instead found baby birds and turtle eggs to be a more accessible diet. Similarly, New Zealand released English stoats in an attempt to control another imported species, the rabbit, but the stoats proceeded to decimate native birds, including the national bird, the kiwi. In the 1930s and 1940s, the U.S. Soil Conservation Service paid farmers to plant kudzu, a plant native to Japan, to prevent erosion and provide ornamental hedges, but it grew so rapidly that it has literally buried some forests in southern states.

These and many other cases in which release of nonnative species into a new environment created unintended consequences should make us pause before introducing genetically modified CRISPR gene drive mosquitos into malaria-stricken parts of Africa. What's more, even if gene drive worked as intended for mosquito control, it would inevitably spark interest in extending the technology to eradicate invasive species such as kudzu, black rats, zebra mussels, and the giant African snail. Each project would be enticing, but each would bring its own list of potential pitfalls that would need careful consideration.

This leaves us with a conundrum. Wiping out the *Anopheles* mosquito would largely wipe out malaria, preventing millions of sicknesses and half-a-million deaths per year, mostly of children. This makes

CRISPR gene drive extremely attractive. It seems as if the environment would hardly miss *Anopheles*, as there are many other mosquito species that would continue to thrive. Yet, yet, yet ... it's the things we don't know that should keep us from pulling the trigger. That's why the U.S. National Academies, in a yearlong study completed in 2016, concluded that more research is needed before CRISPR gene drive is implemented.

A CRISPR WORLD

While the jury is still out on CRISPR gene drive, it's worth considering less invasive applications of this gene-editing technology that could help the planet. It's abundantly clear that Earth's climate is changing. Between 2014 and 2023, we experienced eight of the hottest years on record. Persistent drought in the western United States has caused Lake Mead and Lake Powell, the country's two largest water-storage reservoirs, to drop to historically low levels. Rising ocean temperatures are causing coral reefs to undergo bleaching, as the coral release the algae that provide food for them and other marine creatures. Healthy coral reefs are not just beautiful but also essential to the world's food supply as the home of one quarter of all marine life.

What does this have to do with CRISPR gene-editing? Modification of genes or insertion of new genes into a genome could provide a pathway to making an organism better able to survive and thrive in a hotter, drier world. Sure, given enough time, through Darwinian evolution coral reefs, crop plants, and a host of other at-risk species would probably adapt to climate change without any help. But climate change is happening so quickly, random mutations and natural selection can't keep up. If these organisms are to avoid extinction, they may need a prod. So scientists are using CRISPR to make crop plants more resistant to heat and drought. Certainly, agrichemical companies were already modifying plant genes before CRISPR was discovered, but the speed, efficiency, and specificity of RNA-guided CRISPR editing have quickly made it the technology of choice. Scientists are even trying to engineer marine coral so the coral can thrive in a warmer ocean without bleaching. Here, the advantages of CRISPR are even more striking. Coral genetics is not a field; there's no preexisting technology, so the fact that CRISPR works in coral—as it does in all organisms that have been tested—makes it the only tool now available for altering the coral genome.

CRISPR is poised to address the climate crisis in another way, by providing better biofuels. When we heat our homes, drive our cars, and generate electricity to recharge our cell phones, we're still mostly reliant on fossil fuels. In the United States in 2022, about 80 percent of the energy we consumed came from oil, coal, and natural gas, and only 20 percent came from hydroelectric, solar, wind, and nuclear energy. The combustion of fossil fuels produces carbon dioxide, a powerful greenhouse gas that warms our planet. But if we used plants or algae to produce biofuels, then it would be closer to a break-even proposition: through photosynthesis, plants remove carbon dioxide from the atmosphere while they are alive, compensating for the carbon dioxide that's released when the fuel they make is burned.

This calculus led to initial enthusiasm—and attractive government incentives—to convert corn to the biofuel ethanol, which is then mixed with gasoline in an attempt to make the gasoline "greener." But there are problems with this model, including the diversion of corn away from the food supply and the fact that the energy required to grow and process corn largely cancels out the benefit of its carbon fixation. Algae, conversely, produce more than 20 times the fuel per acre that corn produces, can be grown on land that's of no use for farming, and use brine water instead of the ever-scarcer freshwater required by corn. This is where CRISPR genome-editing enters the picture. Algae have no evolutionary incentive to be super ethanol-producers, but their genomes can be reengineered to greatly enhance their production of this biofuel.

Methane, the same gas used in gas furnaces and gas ranges, is a powerful promoter of global warming if it's released into the atmosphere rather than burned. A major source of methane release always brings a chuckle: cow belches and farts. An amazing 40 percent of annual methane release comes from grazing animals—or rather, the bacteria in their digestive systems. It seems quite likely that CRISPR gene-editing could be used to divert these bacteria from releasing methane to fixing carbon in a safe molecule, such as a sugar or a fat.

Another way to reduce greenhouse gases in the atmosphere is to ramp up what plants are always doing: using photosynthesis to convert carbon dioxide into sugar plus oxygen. Jennifer Doudna's Innovative Genomics Institute at UC Berkeley has undertaken a large program to improve the inherent capacity of plants and soil bacteria to remove carbon dioxide from the air and store the carbon. They're using CRISPR genome-editing to improve the amount of photosynthesis a plant can perform and to increase the carbon-storing capacity of the root systems. Instead of simply remediating the consequences of climate change, these approaches have the potential to actually reverse the process itself. Will CRISPR, powered by its guide RNA, help rescue the planet? Time will tell.

Scientists were amazed to find that bacteria had harbored the CRISPR system for eons before they discovered it—and even more amazed to see how it could be reengineered to cut or modify specific DNA sequences in the human genome. Yet, if we look at it from another angle, we see that RNA is up to its same old tricks. In every iteration of CRISPR, a guide RNA uses the power of nucleic acid base-pairing to bring the editing machinery to a specific site in a complex genome. Then an associated protein—whether it be catalytically active Cas9, Dead Cas9, or some other member of the family—performs some action on that DNA sequence.

We've encountered this principle before, in both RNA interference and telomerase: RNA provides the guide, and, once the RNA finds the site of action, a protein enzyme performs a catalytic act. Part of the reason why CRISPR gene-editing exploded so quickly is that it conforms to the pattern established by these earlier RNA breakthroughs. Scientists were given a revolutionary new pair of scissors, but they already knew how to run with them.

Epilogue

THE FUTURE OF RNA

Cosmologists are intensely focused on understanding the nature of dark matter and dark energy. Since the Big Bang, the universe has been expanding. But over time, as stars exert their gravitational force on each other, the rate of cosmic expansion should be slowing down. Instead, astronomers have found that the expansion of the universe is speeding up. To explain this anomalous motion of stars and galaxies, they propose that only 5 percent of the content of the universe is visible to us. The rest is unseen: 27 percent is dark matter, and 68 percent is dark energy.

Although we can't see it, we know that dark matter *matters*, at least when it comes to thinking about astronomy. But what about biology? As it turns out, our genome also largely consists of "dark matter." The coding regions of all the human genes that specify proteins make up only about 2 percent of our genome. When we add the introns that interrupt those coding regions—the sequences that are spliced out after the DNA is transcribed into the precursors to mRNA—we account for another 24 percent. That leaves about three-quarters of the genome that is "dark matter." For decades this 75 percent was dismissed as "junk DNA" because whatever function it had, if any, was invisible to us.

But as the technologies for sequencing RNA have improved, scientists have discovered that most of this dark-matter DNA is in fact transcribed into RNA. Some portion of this DNA is copied into RNA in the brain, other portions in muscle, or in the heart, or in the sex organs. It's only when we add up the RNAs made in all the tissues of the body that we see the true diversity of human RNAs. The total number of RNAs made from DNA's "dark matter" has been estimated to be several hundred thousand. These are not messenger RNAs, but rather noncoding RNAs—the same general category as ribosomal RNA, transfer RNA, telomerase RNA, and microRNAs. But what they're doing is still, for the most part, a mystery.

The RNAs that emerge from this dark matter are called *long noncoding RNAs* (lncRNAs). While they are particularly numerous in humans, they are also abundant in other mammals, including the laboratory mouse. In a few cases, they clearly have a biological function. For example, a lncRNA called Firre contributes to the normal development of blood cells in mice; an overabundance of Firre prevents mice from fending off bacterial infections, as their innate immune response fails. Another lncRNA, called Tug1, is essential for male mice to be fertile. But such verified functions are few and far between. The function of most lncRNAs remains unknown.

As a result, many scientists do not share my enthusiasm for these RNAs. They think that RNA polymerase, the enzyme that synthesizes RNA from DNA, makes mistakes and sometimes copies junk DNA into junk RNA. A more scholarly description of such RNAs might explain them away as "transcriptional noise"—the idea being, again, that RNA polymerase isn't perfect. It sometimes sits down on the wrong piece of DNA and copies it into RNA, and that RNA may have no function. I readily admit that some of the lncRNAs may in fact be noise, bereft of function, signifying nothing.

However, I'll point out that there was a time in the not-too-distant past when telomerase RNA and microRNAs and catalytic RNAs weren't understood. They hadn't been assigned any function. They, too, could have been dismissed as "noise" or "junk." But now hundreds of research scientists go to annual conferences to talk about these RNAs, and biotech companies are trying to use them to develop the next generation of pharmaceuticals. Certainly one lesson we've learned from the story of RNA is never to underestimate its power. Thus, these lncRNAs are likely to provide abundant material for future chapters in the book of RNA.

Humans and mice are not the only places where yet-to-be-discovered RNAs are likely hiding. The world is full of critters whose biology remains unexplored. Consider the RNA discoveries recounted in this book—and the lowly organisms that gave rise to them. Investigating the microscopic pond scum inhabitant *Tetrahymena*, for instance, led to the discovery not only of catalytic RNA but also of telomerase, which gave us insights into key processes behind cancer and aging in humans. *Euplotes* gave us TERT, the protein partner of telomerase RNA, and the gene for the corresponding human TERT turned out to be the third most frequently mutated gene in all of cancer. Studying how bacteria defend themselves against viruses gave us the broad genome-editing power of CRISPR. An *E. coli* virus, T7, contributed the RNA polymerase that makes the lifesaving mRNA vaccines. Worms unlocked the secrets of an entirely unexpected mode of gene regulation—RNA interference—that was working away in humans, as well, but had gone unnoticed.

In none of these cases did we researchers know exactly where our investigations would lead. In most cases, we were not expecting that our work would eventually produce a cure for a disease or a new tool for biotechnology. We were instead driven by curiosity, trying to understand fundamental biological phenomena. The esoteric creatures we chose to study exaggerated one of these phenomena in some way, making a complex subject more accessible, or offered some other practical advantage in the lab. Then, because we believed in evolution—that all living things are related through the great tree of life—we knew that whatever we found in the obscure organism would have implications for other organisms as well.

Yet, while RNA science—and, through it, medicine and biotechnology—has benefited enormously from studying little-known critters over the past half century, funding agencies are downsizing their support for such research. In fact, funding for any sort of fundamental, curiosity-driven research has declined in recent decades. The National Institutes of Health—by far the largest supporter of biomedical research in the world, with an annual research budget of more than $30 billion—has cut off much of its funding for studies of animals such as *Tetrahymena*. The NIH now emphasizes disease-oriented research using human cells or patients or mice, and to a small extent research using yeast, worms, and fruit flies. I know of scientists who have left biology research in frustration or retired early because the "pond scum" they studied was considered too distant from humans to inform us about human biology.

It's easy to understand the attraction of funding disease-oriented research. It's much easier for a member of Congress or a government official to tout increased funding for breast cancer or prostate cancer than for pond scum, fungi, and worms. But the story of RNA illustrates that many of our most promising new drugs and therapies have come out of research that was driven by scientific curiosity alone. I believe that with a more balanced portfolio of research priorities, we can tackle both specific diseases and the fundamental questions posed by basic science at the same time. We should do so with the humility to recognize that the next major medical breakthrough might again come from an unlikely source.

Nature isn't the only place where we can discover new RNAs. Following the pioneering work of researchers such as Sol Spiegelman, scientists have been able to fast-track the process of evolution in the test tube, revealing novel potentials of RNA beyond those that occur naturally.

One class of such RNAs, called *aptamers*, fold up into shapes that bind specifically to a given protein or a small molecule, much like a protein antibody engages its target. In 1990, Craig Tuerk and Larry Gold at the University of Colorado Boulder and Andy Ellington and Jack

Szostak at Harvard independently demonstrated that you could make a vast collection of different RNA sequences, then isolate the exceedingly rare one that would bind to a chosen target. Their experimental design takes advantage of the dual nature of RNA as both a functional and an informational molecule.

These test-tube evolution methods start with more than a trillion different RNA sequences. The ones that can bind to the target—perhaps a protein from the coat of a virus—are captured, while the "loser" molecules that can't bind are washed away. The challenge is that almost all of the trillion RNAs are losers that can't perform this task; it's a rare RNA that just happens to fold into a structure that does the job. So how do you find this one-in-a-trillion winner? This is where RNA's informational properties come in. You use a polymerase to copy the winning RNA, over and over, in a process called the polymerase chain reaction (PCR). Soon the lonely winner RNA is surrounded by millions of identical copies, so that its sequence of A, G, C, and U bases can be determined.

RNA is so apt at folding into a myriad of shapes that Larry Gold founded a company, SomaLogic, that has turned these aptamers into a diagnostic platform. They have made 7,000 aptamers, each recognizing and measuring the amount of a single human protein in a drop of blood. Researchers around the world have used these aptamers to follow changes in the abundance of particular proteins that can give advance warning of the progression of heart disease and various cancers. Pharmaceutical companies are using information gleaned from these aptamers to identify new proteins that they could target to treat specific diseases.

Because aptamers that bind a wide variety of molecules can be quickly identified, they are also being developed as biosensors. For example, RNA aptamers that bind mercury or lead can be used to identify these toxic elements in environmental samples. In the agricultural industry, aptamers that bind molecules on the surface of pathogenic *E. coli* or that bind viruses or that bind antibiotics can be used to monitor fresh produce or ground meats for the presence of such unhealthy contaminants.

Finally, the ability of RNA aptamers to bind specific proteins—an activity not unlike that of antibodies—gives them therapeutic potential as well. The first therapeutic aptamer was approved by the U.S. Food and Drug Administration in 2004 to treat age-related macular degeneration, a leading cause of vision loss. Macugen, a 27-nucleotide RNA aptamer, binds and thereby inactivates a growth factor called VEGF that triggers the growth of blood vessels across the eye's retina. Injected directly into the eye every six weeks, Macugen improved vision in about one-third of the subjects in the clinical trial. Although it was soon replaced by more effective VEGF inhibitors, it proved the point that aptamers could be useful as therapeutics.

Also in 1990, Jerry Joyce at the Salk Institute was the first to use test-tube evolution to find artificial RNAs that could perform novel catalytic feats—ribozymes, but unlike any found in nature. The general protocol is similar to what Larry Gold and Jack Szostak used to find aptamers: a vast collection of random RNA sequences is challenged with a task—in one case, cleaving a DNA molecule at a specific spot—and the rare molecules that can do the job are collected and amplified. Other scientists have used this approach to find artificial ribozymes that can construct their own nucleotide building blocks, act as RNA polymerases, or attach amino acids to RNA. These novel ribozymes reveal that RNA's powers of catalysis are even more versatile than previously imagined, supporting the view that life on Earth could have emerged in an "RNA world."

Thus the future chapters of the book of RNA will be drawn from both natural and artificial sources. In the first case, we have a vast reserve of unexplored RNAs lurking in diverse critters and in the unfathomable depths of the human genome. In the second, we have found ways to evolve novel RNAs capable of tricks that—as far as we know—do not occur in nature. The lesson is one we've learned before: never underestimate RNA.

In the course of writing this book, I've striven to keep RNA as the protagonist of the many dramas I've witnessed. After all, RNA is the consummate catalyst. RNA can catalyze its own rearrangement when it undergoes self-splicing, and it catalyzes the variations in RNA splicing that allow humans to get so much action out of our rather limited genome. RNA catalyzes the construction of all the proteins that form the structures and the enzymes of all the cells in every human being and in every other creature on Earth. RNA teams up with proteins to catalyze the extension of our chromosome ends, our telomeres, allowing human embryos to develop and stem cells—as well as, unfortunately, tumor cells—to keep on dividing. RNA teams up with proteins to catalyze the silencing of gene expression in a process called RNA interference. Another RNA-protein team called CRISPR catalyzes the destruction of bacterial viruses and provides unprecedented power for editing the fundamental code of our DNA. RNA catalyzes the potency of human viruses, but at the same time—packaged in lipid vessels—it catalyzes protection from these very viruses in the form of mRNA vaccines. RNA has been catalyzing life from its very origins, working its magic as an enzyme while simultaneously serving as an informational molecule—or so we think.

But as much as I've worked to keep the spotlight on RNA, its story intersects with my own journey, so I have sometimes strayed from my role as narrator and stepped onstage for a bit. When I finished my PhD and my postdoctoral studies, working exclusively with DNA, I had no idea that RNA would so soon dominate all my thoughts and efforts. This transition from DNA scientist to RNA scientist was not mine alone; it was the path taken by many in the early days of the field. During this same time, RNA was stepping out of the shadows—no longer just a tool at the service of DNA, but a wondrous molecule of limitless possibilities. I feel privileged to have been able to ride shotgun with RNA at every twist of the journey.

ACKNOWLEDGMENTS

I decided to write this book in June of 2021. I was confident that I could translate complex scientific concepts to the general public because, after all, I had taught freshman chemistry to several thousand undergraduate students at the University of Colorado. How much more difficult could it be to explain RNA to a slightly more general audience? It turned out that my confidence was in fact overconfidence. Unlike my chemistry students, who have recently taken science courses in high school, their parents are for the most part long removed from any study of science. Writing a book to engage this broader group of inquisitive nonscientists ended up being a two-year struggle.

I could not have stuck with the writing without help from many corners. Fortuitously, I met Steve Heyman, who had the same modest level of biochemical knowledge as my intended audience. The suggestions he made on every version of every chapter were critical to the final product. Jessica Yao, my editor at W. W. Norton, was fearless about questioning almost every sentence. Was the organization optimal, and were my explanations too technical? I would have pushed back more often, except that I could see that she was almost always right. I was also fortunate to find Zovinar "Zovi" Khrimian, whose meticulous ink drawings helped bring complex concepts to light. Her illustrations portray RNA in a consistent manner that we hope will simplify ideas for the reader. Finally, I thank my agents, Peter and Amy Bernstein, for their enthusiasm and for finding such a good publisher.

Jennifer Doudna and John Inglis helped at the beginning with encouragement for the project and advice about publishing. As I began writing, I cornered many scientists for in-depth interviews or conversations, and almost without exception they were excited about helping bring RNA to the people. These included John Abelson, Dana Carroll, Phil Felgner, Elfriede Gamow, Cecilia Guerrier-Takada, Christine Guthrie, Franklin Huang, Melissa Moore, Harry Noller, Norm Pace, Dan Rokhsar, Joan Steitz, Bruce Sullenger, Eric Westhof, and Meng-Chao Yao. Their recollections added veracity and the occasional great anecdote. I thank my colleague Ding Xue and his postdoctoral fellow Joyita Bhadra for discussion and a hands-on demonstration of injecting nematode worms. I relied on Paul Rothman's expertise to check my description of some medical concepts with which I was unfamiliar.

I am grateful to the 100 graduate students and postdoctoral fellows and similar number of undergraduate students who've trained in my lab over these decades. Although the book mentions only a few of you by name, the joys and frustrations of research that we've experienced together have shaped the way that I explain the process to the general reader. So you are all there, behind the pages, contributing to the tone of the book, as are my longtime research associate, Art Zaug, my colleague Olke Uhlenbeck, and my friend Tom Mann.

Special thanks to my wife, Carol, for being a sounding board when needed and for generously tolerating my long retreats to my office cave every night. Thanks to my daughters and sons-in-law for understanding why I sometimes had to skip skiing or other activities. And thank you Skyler for the joy of life you expressed in all your comments when I walked you to kindergarten on Wednesday mornings. You, Bradley, and Benjamin are wonderful reminders of the magic of human development—due in no small part to The Catalyst, RNA.

GLOSSARY

adaptive immunity—A process of protection from infection that is specific to a particular pathogen or other foreign substance. In contrast to the innate immune system, the adaptive (acquired) immune system requires prior exposure to the pathogen, either through infection or vaccination, before it stands ready to respond to a new exposure. *See also* **innate immunity**.

alpha-globin—One of two types of **protein** chains that make up hemoglobin, the oxygen-carrying protein found in red blood cells. Hemoglobin has two alpha-globin and two beta-globin subunits, which are similar but not identical protein chains. *See also* **beta-globin**.

alternative mRNA splicing—The process by which the use of different splice sites produces different mRNA transcripts, resulting in two or more **proteins** being encoded by a single gene.

amino acid—One of the building blocks of **proteins**. There are 20 common amino acids, each specified by between one and six mRNA **codons**.

amino acid sequence—The specific order of building blocks (amino acids) along a **protein** chain, which determines how the protein folds in three dimensions and attains a particular function—such as digesting food in the stomach, making muscles move, or carrying oxygen in the bloodstream.

GLOSSARY

anticodon—The three **bases** of a transfer RNA molecule that base-pair with an mRNA codon.

antisense RNA—A tool for inhibiting the function of a gene by using an RNA that is complementary (antisense) to a target mRNA.

aptamer—An artificial nucleic acid molecule selected to bind specifically to a **protein** or a small **molecule**; from the Latin *aptus*, "to fit."

Argonaute—A **protein** required for **RNA interference** that binds **siRNAs** and **microRNAs** and uses them as guides to degrade or inhibit the **translation** of a group of mRNAs.

B cells—**Lymphocytes** that protect an animal from infection by producing antibodies that bind to and thereby inhibit invading viruses and bacteria.

bacteriophage—A DNA or RNA virus that infects bacteria.

base pair—An interaction involving G (guanine) pairing with C (cytosine) or A (adenine) pairing with T (thymine) or U (uracil). Base pairs form the rungs in the twisted ladder of the DNA double helix. RNA structure depends on base pairs that form within a single strand.

bases—Chemical units present at every position along a DNA or RNA chain that are the fundamental units of nucleic acid information. Three of the bases (A, G, C) are identical in DNA and RNA, while the fourth is T in DNA and U in RNA. *See also* **nucleotides**.

beta-globin—One of two types of **protein** chains that make up hemoglobin, the oxygen-carrying protein found in red blood cells. Hemoglobin has two alpha-globin and two beta-globin subunits. *See also* **alpha-globin**.

capsid—A protective **protein** shell that shields viral RNA from the perils of living tissues—such as **ribonucleases**—as it guides the RNA to and into its target cell.

cell membrane—The two-layered envelope of **lipid** molecules that forms the protective outer layer of many living cells.

chromosome—A discrete package of DNA and **proteins** that carries the genetic information of an organism. Before the DNA undergoes **replication**, each chromosome contains one DNA double helix.

GLOSSARY

codon—A group of three **nucleotides** in an mRNA that specifies a particular **amino acid** in the **protein** product.

complementary—Matching; for example, in RNA base-pairing, G is complementary to C.

CRISPR (clustered regularly interspaced short palindromic repeats)—A DNA-cleaving system, powered by a guide RNA (which targets the section of **DNA sequence** to be cut) and the Cas9 (CRISPR-associated 9) **protein** (which does the cutting). CRISPR was discovered as a natural process in bacteria, which use it to stave off attacks by viruses known as bacteriophage.

Dicer—A **protein enzyme** that chops long double-stranded RNAs into siRNAs and chops precursors to **microRNAs** into mature microRNAs.

DNA sequence—The specific order of building blocks (**nucleotides**) along a strand of DNA.

DNA vaccine—Uses the gene that encodes a viral or bacterial protein to train the **adaptive immune** system to be on the lookout for that pathogen. DNA vaccines rely on the human recipient to copy the DNA into mRNA and then into **protein.**

donor template—A DNA molecule used in **CRISPR** gene-editing that provides a template for **homologous recombination** after the DNA is cleaved. Sequences of the donor template become part of the repaired **chromosome**.

enzyme—Substance in living cells that speeds up a biochemical reaction needed for life without being consumed in the reaction. Enzymes, which are usually **proteins**, make our heart beat, break down the food in our stomach, and synthesize all of the structures that hold our cells together.

eukaryotes—Organisms, from algae to humans, that form a cell nucleus to sequester their DNA.

exocytosis—The process by which mature virus particles exit an infected cell by hitchhiking on a pathway that the cell has developed to export some of its own **proteins.**

familial disease—One that runs within a family because a disease-causing mutation is passed from parents to children.

fermentation—**Enzyme**-catalyzed conversion of sugar or starch to alcohol in yeast and other microorganisms.

gel electrophoresis—A research method used to separate macromolecules such as DNA and RNA. The gel is a slab of wiggly Jell-O–like material, and when an electric field is applied across it—negative electrode at the top, positive electrode at the bottom—negatively charged DNA or RNA molecules that start at the top are pushed through the gel, being separated according to their size.

gene therapy—Use of DNA to treat or prevent a disease, most commonly by giving a person a new copy of a healthy gene to compensate for a defective gene.

genome—The totality of the DNA in an organism. The human genome, bundled into 23 **chromosomes**, has about 3 billion **bases**.

germline cells—Sex cells (sperm and egg) and cells that give rise to sex cells, which differ from **somatic cells** in that their genetic information is passed on to the organism's offspring.

homologous recombination—A process for repair of double-stranded breaks in DNA that requires an intact version of an identical or similar (homologous) sequence to guide the repair. An intermediate step in this repair process involves exchange of genetic information between the two versions of the DNA (recombination).

innate immunity—A process of protection from infection that differs from **adaptive immunity** in that it is not specific to a particular pathogen. The innate immune system recognizes conserved features of pathogens and is quickly activated to destroy invaders.

introns—Stretches of DNA that do not code for **proteins** and are spliced out of the RNA after **transcription**. In the case of rRNA and tRNA introns, they are stretches that interrupt the sequence of the mature, functional molecules, and again they must be spliced out. The same term is used for both the DNA and RNA elements.

large subunit of the ribosome—The portion of the protein-synthesizing machine containing the active site for the **peptidyl transfer** reaction that connects **amino acids** to form **proteins**. In *E. coli* it consists of two **rRNAs** (2,904 and 120 nucleotides) and some 33 proteins.

lipid envelope—For some RNA viruses, the **capsid** provides enough of a "space capsule" to protect the RNA and deliver it to its destination. In other viruses, however, the capsid is surrounded by yet another layer—an envelope made of fatty **lipid** molecules.

lipid nanoparticle (LNP)—A drug-delivery system made of fatty **lipid** molecules that allows an mRNA-based vaccine or therapeutic to bypass cellular defenses.

lipids—Fats, waxes, and oils that do not dissolve much in water. Lipids make up the membranes that surround animal cells and the outer layer of enveloped viruses.

liposome—A vehicle for delivering DNA or RNA into cells that has a double layer of **lipids** on its exterior and encapsulates the nucleic acid in its interior.

long noncoding RNAs (lncRNAs)—RNA molecules larger than 200 **nucleotides** that do not encode **proteins**. The functions of most lncRNAs are unknown.

lymphocytes—White blood cells that are part of the immune system. They consist mostly of **B cells** and **T cells**.

messenger RNA (mRNA)—An RNA containing a string of **codons** that instructs the synthesis of a particular **protein**. Human mRNAs carry the message from the DNA in the cell nucleus to the **ribosomes** in the cytoplasm.

microRNA—A very small RNA that binds to **mRNAs** in the cell cytoplasm and cleaves them or inhibits their **translation**, thereby interfering with gene expression at the RNA level. MicroRNAs are initially made as larger RNAs that **base-pair** within themselves to form long double-stranded segments, which are then cut down to size by **Dicer**. They contribute to processes including the development of arms and legs, the formation of heart muscle, the production of blood cells, particularly of immune cells, and placental development and pregnancy. When microRNAs are perturbed, they contribute to many diseases.

minichromosome—A small DNA element that is replicated independently of the chromosomal DNA in the same cell. A minichromosome is a natural version of a **plasmid**. The *Tetrahymena* **ribosomal RNA** genes occur on minichromosomes.

molecule—A group of atoms held together in a specific way by chemical bonds. Water, cane sugar, and carbon dioxide are molecules. DNA, RNA, and proteins are also molecules, which because of their size are called macromolecules.

mRNA vaccine—A product that triggers the immune system to be on the lookout for a particular pathogen by use of an **mRNA** that encodes a surface **protein** of the pathogen. Human cells translate the mRNA into the protein characteristic of the pathogen, which then instructs the adaptive immune system. *See also* **adaptive immunity**.

negative (–) strand RNA virus—The viral RNA enters the host as the complement to **mRNA**. These viruses bring along their own copying **enzyme** or **replicase**, which copies the (–) strands into (+) strands that serve as mRNAs. All the flu (influenza) viruses are (–) strand, as are respiratory syncytial virus (RSV), rabies virus, and Ebola virus.

noncoding RNAs—A group of RNAs that play critical roles in cellular biology even though they do not code for proteins. These include **ribosomal RNAs**, **transfer RNAs**, **telomerase RNA**, **small nuclear RNAs**, and **ribozymes**.

non-homologous end joining (NHEJ)—A process for repair of double-stranded breaks in DNA in which the broken ends are directly tied together without the aid of an intact version of the same (homologous) sequence to guide the repair. NHEJ often leaves small insertions or deletions of nucleotides at the site of repair. *See also* **homologous recombination**.

nucleotides—Fundamental building blocks of DNA or RNA, each consisting of a phosphate group, a sugar (deoxyribose or ribose), and a **base** (A, G, C, and either T or U; DNA contains T, RNA contains U). The information content of nucleotides is identical to that of their bases.

oncogene—A mutated form of a normal human gene that drives abnormal cell division and causes cancer.

peptidyl transfer—The chemical reaction that joins two **amino acids** together, building a **protein** chain.

phage—Short for **bacteriophage**, a virus that infects bacteria.

plasmid—An artificial DNA element that replicates independently of a cell's **chromosome**(s), typically as a small circular DNA. Plasmids are used in research for the isolation and manipulation of genes.

positive (+) strand RNA virus—The viral RNA enters the host ready to serve as **mRNA** coding for viral **proteins**. The viral mRNAs hijack the host cell's **ribosomes** to produce their poison proteins. Positive (+) strand RNA viruses include poliovirus, dengue virus, hepatitis A and C viruses, SARS-CoV-2, and rhinovirus, the last of these causing the common cold.

precursor—In the case of RNA, the form of an RNA initially transcribed from a gene before it is trimmed, spliced, or otherwise modified and ready to do its job.

proteins—Strings of **amino acids** that fold into specific shapes to perform diverse functions. In animals, some proteins form structures such as muscle fibers, skin, and hair. Some of them act as **enzymes**, breaking down the food we eat into its constituent components and then recycling these pieces to build up new cellular machines. Some of them punch holes in the envelopes that encase our cells, selectively allowing some salts or nutrients to enter the cell and expelling others. Some act as signaling molecules, receiving information from the outside world and activating internal processes accordingly. Some are antibodies, which protect us from foreign invaders such as viruses.

pseudoU—A modified version of uracil that is found naturally in specific locations in some **tRNA** molecules. It still forms A-U base pairs but is no longer recognized by the innate immune system. *See also* **innate immunity**.

receptor—A **protein** that binds specifically to a **molecule** such as a different protein or a hormone or an odorant. Receptors allow cells to respond to their external environment.

replicase—A general term for an **enzyme** that copies or reproduces a nucleic acid into more nucleic acids of the same type. DNA polymerases and viral RNA-dependent RNA polymerases are examples of replicases.

replication—The process of producing identical or near-identical copies of a single nucleic acid molecule. In DNA replication, one double helix is copied into two double helices. In RNA replication, found mostly in viruses, one single-stranded RNA is copied into multiple progeny strands.

reverse transcriptase—An **enzyme** that transcribes single-stranded RNA into DNA, a reversal of the normal DNA-to-RNA **transcription**.

ribonuclease (RNase)—An **enzyme** that cuts RNA.

ribonuclease P (RNase P)—The **enzyme** that cleaves the leader sequence from a **tRNA** precursor, forming the mature, active tRNA. In bacteria it consists of a catalytic RNA moiety (a **ribozyme**) and a supporting **protein** moiety. *See also* **precursor**.

ribosomal RNA (rRNA)—A **noncoding RNA** that is essential for **ribosome** function. Bacterial ribosomes contain three ribosomal RNAs, whereas eukaryotic ribosomes contain four.

ribosomes—The cellular factories for making **proteins**. Ribosomes consist of RNA (**rRNA**) and ribosomal proteins.

ribozyme—A ribonucleic acid with enzymatic activity.

RNA interference (RNAi)—A regulatory process in nature that allows organisms to reduce the activity of groups of mRNAs after they are transcribed. RNAi has also been repurposed into a therapy to treat rare genetic diseases. *See also* **microRNA** and **siRNA**.

RNA polymerase—The **enzyme** that copies information from DNA into RNA.

RNA processing—Steps that must occur after the initial transcription of a **precursor** RNA in order to form a functional RNA. These include **RNA splicing** and cutting off unneeded sequences as well as adding bases not encoded by the DNA.

RNA sequence—The specific order of building blocks (**nucleotides**) along a strand of RNA.

RNA splicing—The biochemical process by which interrupting sequences (or **introns**) of RNA **precursors** are cut out and flanking sequences joined together.

senescence—An end stage in the aging of a cell, where the cell continues to live and changes its metabolism but no longer divides.

single-guide RNA—An RNA engineered for **CRISPR** gene-editing that combines the guide RNA and **tracrRNA**.

small interfering RNA (siRNA)—An artificial 23-nucleotide double-stranded RNA, one strand of which is complementary to a target mRNA. Through the **RNA interference** pathway, the siRNA directs the cleavage and inactivation of the target mRNA.

small nuclear RNA (snRNA)—Stable, abundant noncoding RNAs that exist as complexes with specific **proteins**. U1 and U2 snRNAs mark the sites of splicing along the mRNA precursor, and U2, U5, and U6 snRNAs are directly involved in the catalysis of mRNA splicing. U4 snRNA keeps U6 snRNA in check during early stages of splicing until it is displaced. U3 snRNA is not involved in mRNA splicing but rather in the maturation of **rRNA**.

small subunit of the ribosome—The portion of the **protein**-synthesizing machine that is the first to assemble with the mRNA. It directs the decoding of the mRNA and the binding of **tRNAs**. In *E. coli* it consists of the second-largest of the three **rRNAs** and 22 proteins.

somatic cell—All body cells other than **germline cells**. These include skin, muscle, liver, blood, and brain cells. In humans they are diploid (two sets of **chromosomes**) and do not pass on any acquired mutations to the organism's offspring.

spinal muscular atrophy (SMA)—A deadly neurodegenerative disease caused by mutations in the **SMN1** gene.

splicing—*See* **RNA splicing**.

splicing factor—Any of a large number of **proteins** that facilitate and regulate mRNA splicing. These factors are distinct from the **snRNA**-protein complexes that directly catalyze splicing.

sporadic disease—One that arises spontaneously without any family history of the disease. Sporadic diseases can still have a genetic cause due to a mutation in a **somatic cell**.

survival of motor neuron (SMN1 and SMN2)—The SMN1 gene encodes a **protein** involved in assembly of **snRNA**-protein complexes. When mutated, it causes **SMA**. The SMN2 gene can be activated to rescue the loss of mutant SMN1.

T cells—**Lymphocytes** that protect an animal by destroying any of its cells that have been infected by a pathogen or become cancerous.

GLOSSARY

telomerase—A molecular machine that enables eukaryotic cells to keep dividing by adding protective DNA sequences to the ends of **chromosomes (telomeres)**. Telomerase consists of an RNA molecule and **proteins**, including **TERT**.

telomerase reverse transcriptase (TERT)—The protein partner of **telomerase** RNA that contains the catalytic center for telomeric DNA synthesis.

telomere—The end of a chromosome, consisting of a repeated DNA sequence and associated protective **proteins**. Telomere length serves as a clock to measure the number of cell divisions that a **somatic cell** can undergo.

tracrRNA—"trans-activating **CRISPR** RNA," a naturally occurring bacterial RNA that pairs with the guide RNA and secures it to the Cas9 protein.

transcription—The process by which DNA is copied into RNA.

transfer RNA (tRNA)—A small RNA that transfers the correct **amino acid** to the growing **protein** chain inside the **ribosome**. Each transfer RNA recognizes an mRNA **codon** corresponding to its amino acid.

translation—The process of reading out the mRNA code to synthesize a **protein**. It is carried out by **ribosomes**.

translocation—The movement of an mRNA **codon** (with its bound **tRNA**) from one site to another within the **ribosome**. This needs to happen each time a codon is read out, in order to make room for the next tRNA to enter.

triplet codon—*See* **codon**.

X-ray crystallography—A technique to determine the structure of molecules, including **proteins** and nucleic acids. It involves shooting a beam of X-rays at a sample, such as a crystal of a protein molecule, collecting images of the diffracted radiation, and calculating what the structure must have been to produce that diffraction.

NOTES

INTRODUCTION: THE AGE OF RNA

3 **have each quadrupled:** Janet M. Sasso, Barbara J. B. Ambrose, Rumiana Tenchov, Ruchira S. Datta, Matthew T. Basel, Robert K. DeLong, and Qiongqiong Angela Zhou, "The Progress and Promise of RNA Medicine—An Arsenal of Targeted Treatments," *Journal of Medical Chemistry* 65, 6975–7015, 2022.

3 **400 RNA-based drugs:** Lin Ning, Mujiexin Liu, Yushu Gou, Yue Yang, Bifang He, and Jian Huang, "Development and Application of Ribonucleic Therapy Strategies Against COVID-19," *International Journal of Biological Sciences* 18, 5070–85, 2022.

3 **new frontiers in RNA research:** Cheryl Barton, "Renewed Interest in RNA-Targeted Therapies—Delivery Remains the Achilles Heel," *Pharma Letter*, January 31, 2023, https://www.thepharmaletter.com/article/renewed-interest-in-rna-targeted-therapies-delivery-remains-the-achilles-heel.

6 **in the endnotes:** For those who want a scholarly step-by-step account of RNA research, I recommend *RNA: Life's Indispensable Molecule* by Jim Darnell (Cold Spring Harbor, NY: Cold Spring Harbor Laboratory Press, 2011) and *RNA: The Epicenter of Genetic Information* by John Mattick and Paulo Amaral (Boca Raton, FL: CRC Press, 2022).

CHAPTER 1: THE MESSENGER

9 **the world's leading proponent:** Paul Halpern, *Flashes of Creation: George Gamow, Fred Hoyle, and the Great Big Bang Debate* (New York: Basic Books, 2021), 2.

9 **"another Heisenberg"**: Karl Hufbauer, *George Gamow, 1904–1968: A Biographical Memoir* (Washington, DC: National Academy of Sciences, 2009), 9.
9 **"giant imp"**: James Watson, *Genes, Girls, and Gamow* (Oxford: Oxford University Press, 2003), xxiv.
11 **"exact sciences"**: Hufbauer, *George Gamow*, 25.
11 **"had so many whimsical qualities"**: Watson, *Genes, Girls, and Gamow*, 24.
12 **amino acids along its chain:** F. Sanger and H. Tuppy, "The Amino-Acid Sequence in the Phenylalanyl Chain of Insulin. 1. The Identification of Lower Peptides from Partial Hydrolysates," *Biochemical Journal* 49, 463–81, 1951.
13 **focus their code-breaking attention on RNA:** Author's interview with Dan Rokhsar, Boulder, Colorado, October 4, 2023.
13 **proteins are made *outside* the nucleus:** Jean Brachet, "La détection histochimique et le microdosage des acides pentose-nucléiques," *Enzymologia* 10, 87–96, 1942; Torbjörn Caspersson, "The Relation Between Nucleic Acid and Protein Synthesis," *Symposia of the Society for Experimental Biology* 1, 129–51, 1947.
13 **chemical compounds without a function:** James Darnell, *RNA: Life's Indispensable Molecule* (Cold Spring Harbor, NY: Cold Spring Harbor Laboratory Press, 2011), 9–10.
13 **heritable changes in bacteria:** Oswald T. Avery, Colin M. MacLeod, and Maclyn McCarty, "Studies on the Chemical Nature of the Substance Inducing Transformation of Pneumococcal Types: Induction of Transformation by a Desoxyribonucleic Acid Fraction Isolated from Pneumococcus Type III," *Journal of Experimental Medicine* 79, 137–58, 1944.
13 **double-helical structure of DNA:** J. D. Watson and F. H. C. Crick, "Molecular Structure of Nucleic Acids: A Structure for Deoxyribose Nucleic Acid," *Nature* 171, 737–38, 1953.
13 **as early as 1947:** André Boivin in Paris first proposed in 1947 that DNA governed the building of RNA, which, in turn, controlled the production of cytoplasmic proteins. See Matthew Cobb, "Who Discovered Messenger RNA?," *Current Biology* 25, R523–R548, 2015.
14 **might code for one amino acid:** Francis Crick, "The Genetic Code," in *What Mad Pursuit: A Personal View of Scientific Discovery* (New York: Basic Books, 1990), 89–101.
14 **"no solution"**: George Gamow, *My World Line: An Informal Biography* (New York: Viking, 1970), 148.
15 **on a visit to Cambridge:** "50th Anniversary of Good Friday Meeting (April 15, 1960)," Cold Spring Harbor Laboratory Press Email News, accessed August 29, 2023, https://www.cshlpress.com/email_news/goodfriday.html.
16 **the stable RNA in the cell:** Kenneth Volkin and Larry Astrachan, "Phosphorus Incorporation in *Escherichia coli* Ribonucleic Acid After Infection with Bacteriophage T2," *Virology* 2, 149–61, 1956.
17 **needed by the phage to do its dirty work:** Sydney Brenner, François Jacob, and

Matthew Meselson, "An Unstable Intermediate Carrying Information from Genes to Ribosomes for Protein Synthesis," *Nature* 190, 576–80, 1961. Jim Watson and his Harvard research group were on the hunt for mRNA at the same time, and their contemporaneous discovery was published back to back with the Brenner et al. paper: François Gros, H. Hiatt, Walter Gilbert, C. G. Kurland, R. W. Risebrough, and J. D. Watson, "Unstable Ribonucleic Acid Revealed by Pulse Labelling of *Escherichia coli*," *Nature* 190, 581–85, 1961.

21 **a clever set of experiments:** Francis H. C. Crick, Leslie Barnett, Sydney Brenner, and Richard Watts-Tobin, "General Nature of the Genetic Code for Proteins," *Nature* 192, 1227–32, 1961.

24 **rat livers:** Elizabeth B. Keller, Paul Zamecnik, and Robert B. Loftfield, "The Role of Microsomes in the Incorporation of Amino Acids into Proteins," *Journal of Histochemistry and Cytochemistry* 2, 378–86, 1954; John W. Littlefield, Elizabeth B. Keller, Jerome Gross, and Paul C. Zamecnik, "Studies of Cytoplasmic Ribonucleoprotein Particles from the Liver of the Rat," *Journal of Biological Chemistry* 217, 111–24, 1955.

24 **UUU encodes phenylalanine:** Marshall W. Nirenberg and J. Heinrich Matthaei, "The Dependence of Cell-Free Protein Synthesis in *E. coli* upon Naturally Occurring or Synthetic Polyribonucleotides," *Proceedings of the National Academy of Sciences USA* 47, 1588–602, 1961. Importantly, the manuscript is careful to state that poly(U) encoding polyphenylalanine did not distinguish whether it was U, UU, UUU, or some other codon that specified Phe.

25 **enzyme to copy the DNA into RNA:** This particular enzyme was *E. coli* RNA polymerase. RNA polymerases are so named because they catalyze the joining of monomers (A, G, C, and U) into polymers of RNA with a sequence determined by the DNA template. There are also DNA polymerases, which perform a similar reaction but use deoxynucleotide monomers of A, G, C, and T.

25 **corresponding amino acid sequences:** S. Nishimura, D. S. Jones, E. Ohtsuka, H. Hayatsu, T. M. Jacob, and H. G. Khorana, "Studies on Polynucleotides: XLVII. The *In Vitro* Synthesis of Homopeptides as Directed by a Ribopolynucleotide Containing a Repeating Trinucleotide Sequence. New Codon Sequences for Lysine, Glutamic Acid and Arginine," *Journal of Molecular Biology* 13, 283–301, 1965.

25 **the 1968 Nobel Prize:** The third recipient of the 1968 Nobel Prize in Physiology or Medicine was Robert W. Holley, for describing the structure of alanine transfer RNA, linking DNA and protein synthesis.

25 **"The solution looks considerably less elegant":** Gamow, *My World Line*, 148.

26 **grand theory of the genetic code:** Matthew Cobb, "60 Years Ago, Francis Crick Changed the Logic of Biology," *PLOS Biology* 15, e2003243, 2017.

26 **mRNA triplet codon:** Francis H. C. Crick, "On Protein Synthesis," *Symposia of the Society for Experimental Biology* 12, 138–63, 1958.

27 **an RNA whose job it was:** Mahlon B. Hoagland, Mary Louise Stephenson, Jesse F.

Scott, Liselotte I. Hecht, and Paul C. Zamecnik, "A Soluble Ribonucleic Acid Intermediate in Protein Synthesis," *Journal of Biological Chemistry* 231, 241–57, 1958.

CHAPTER 2: SPLICE OF LIFE

30 **"Anything found to be true of *E. coli*":** Herbert C. Friedmann, "From 'Butyribacterium' to '*E. coli*': An Essay on Unity in Biochemistry," *Perspectives in Biology and Medicine* 47, 47–66, 2004.

30 **10 times bigger:** James Darnell, *RNA: Life's Indispensable Molecule* (Cold Spring Harbor, NY: Cold Spring Harbor Laboratory Press, 2011), 168–69.

31 **how Phil and the group of Cold Spring scientists solved the riddle:** Susan M. Berget, Claire Moore, and Phillip A. Sharp, "Spliced Segments at the 5' Terminus of Adenovirus 2 Late mRNA," *Proceedings of the National Academy of Sciences USA* 74, 3171–75, 1977; Louise T. Chow, Richard E. Gelinas, Thomas R. Broker, and Richard J. Roberts, "An Amazing Sequence Arrangement at the 5' Ends of Adenovirus 2 Messenger RNA," *Cell* 12, 1–8, 1977.

33 **pruning shears:** Shirley M. Tilghman, David C. Tiemeier, J.G. Seidman, B. Matija Peterlin, Margery Sullivan, Jacob V. Maizel and Philip Leder, "Intervening Sequence of DNA Identified in the Structural Portion of a Mouse Beta-globin Gene," Proceedings of the National Academy of Sciences USA 75, 725–29, 1978.

34 **6,000 protein-coding genes:** A. Goffeau, B. G. Barrell, H. Bussey, R. W. Davis, B. Dujon, H. Feldmann, F. Galibert, J. D. Hoheisel, C. Jacq, M. Johnson, E. J. Louis, H. W. Mewes, Y. Murakami, P. Philippsen, H. Tettelin, and S. G. Oliver, "Life with 6000 Genes," *Science* 274, 546–67, 1996.

35 **24,000 protein-coding genes:** International Human Genome Sequencing Consortium, "Finishing the Euchromatic Sequence of the Human Genome," *Nature* 431, 931–45, 2004.

37 **sugar-coated:** The technical term is *polysaccharide*, which means "many sugars." Many proteins that are secreted from cells have polysaccharides appended to some of their amino acids, and this makes the protein more soluble in aqueous environments just as table sugar is soluble in tea.

37 **The RNA copied from that single gene:** Frederick W. Alt, Alfred L. M. Bothwell, Michael Knapp, Edward Siden, Elizabeth Mather, Marian Koshland, and David Baltimore, "Synthesis of Secreted and Membrane-Bound Immunoglobulin Mu Heavy Chains Is Directed by mRNAs That Differ at Their 3' Ends," *Cell* 20, 293–301, 1980; J. Rogers, P. W. Early, C. Carter, K. Calame, M. Bond, L. Hood, and R. Wall, "Two mRNAs with Different 3' Ends Encode Membrane-Bound and Secreted Forms of Immunoglobulin μ Chain," *Cell* 20, 303–12, 1980; P. W. Early, J. Rogers, M. Davis, K. Calame, M. Bond, R. Wall, and L. Hood, "Two mRNAs Can Be Produced from a Single Immunoglobulin μ Gene by Alternative RNA Processing Pathways," *Cell* 20, 313–19, 1980.

38 **first female graduate student:** Author's interview with Joan Steitz, Boulder, Colorado, March 6, 2022.

38 **"All our wives like being research associates":** Gina Kolata, "Thomas A. Steitz, 78, Dies; Illuminated a Building Block of Life," *New York Times*, October 10, 2018, https://www.nytimes.com/2018/10/10/obituaries/thomas-a-steitz-dead.html.

38 **found in the patients' own cell nuclei:** Michael Rush Lerner and Joan A. Steitz, "Antibodies to Small Nuclear RNAs Complexed with Proteins Are Produced by Patients with Systemic Lupus Erythematosus," *Proceedings of the National Academy of Sciences USA* 76, 5495–99, 1979.

39 **These snRNAs were already known to exist:** Ramachandra Reddy, Tae Suk Ro-Choi, Dale Henning, and Harris Busch, "Primary Sequence of U-1 Nuclear Ribonucleic Acid of Novikoff Hepatoma Ascites Cells," *Journal of Biologic Chemistry* 249, 6486–94, 1974.

39 **a natural propensity to pair up:** Joan A. Steitz and Karen Jakes, "How Ribosomes Select Initiator Regions in mRNA: Base Pair Formation Between the 3' Terminus of 16S rRNA and the mRNA During Initiation of Protein Synthesis in *Escherichia coli*," *Proceedings of the National Academy of Sciences USA* 72, 4734–38, 1975.

40 **maybe U1 was the agent:** Michael R. Lerner, John A. Boyle, Stephen M. Mount, Sandra W. Wolin, and Joan A. Steitz, "Are snRNPs Involved in Splicing?," *Nature* 283, 220–24, 1980. A similar proposal was made at the same time by John Rogers and Randolph Wall, "A Mechanism for RNA Splicing," *Proceedings of the National Academy of Sciences USA* 77, 1877–79, 1980.

40 **collaborative effort found just such behavior:** Richard A. Padgett, Stephen M. Mount, Joan A. Steitz, and Phillip A. Sharp, "Splicing of Messenger RNA Precursors Is Inhibited by Antisera to Small Nuclear Ribonucleoprotein," *Cell* 35, 101–7, 1983.

40 **a protein associated with the U2 snRNA:** Shaoping Wu, Charles M. Romfo, Timothy W. Nilsen, and Michael R. Green, "Functional Recognition of the 3' Splice Site AG by the Splicing Factor U2AF^{35}," *Nature* 402, 832–35, 1999; Diego A. R. Zorio and Thomas Blumenthal, "Both Subunits of U2AF Recognize the 3' Splice Site in *Caenorhabditis elegans*," *Nature* 402, 835–38, 1999; Livia Merendino, Sabine Guth, Daniel Bilbao, Concepción Martínez, and Juan Valcárcel, "Inhibition of msl-2 Splicing by Sex-Lethal Reveals Interaction Between U2AF35 and the 3' Splice Site AG," *Nature* 402, 838–41, 1999.

41 **a 12-year-old Greek Cypriot girl:** Richard A. Spritz, Pudur Jagadeeswaran, Prabhakara V. Choudary, P. Andrew Biro, James T. Elder, Jon K. Deriel, James L. Manley, Malcom L. Gefter, Bernard G. Forget, and Sherman M. Weissman, "Base Substitution in an Intervening Sequence of a Beta+ Thalassemic Human Globin Gene," *Proceedings of the National Academy of Sciences USA* 78, 2455–59, 1981.

41 **caused aberrant mRNA splicing:** Meinrad Busslinger, Nikos Moschonas, and Richard A. Flavell, "Beta+ Thalassemia: Aberrant Splicing Results from a Single Point Mutation in an Intron," *Cell* 27, 289–98, 1981.

41 *survival of motor neuron* **gene:** Livio Pellizzoni, Bernard Charroux, and Gideon Dreyfuss, "SMN Mutants of Spinal Muscular Atrophy Patients Are Defective in Binding to snRNP Proteins," *Proceedings of the National Academy of Sciences USA* 96, 11167–72, 1999.

42 **a protein that helps snRNAs assemble:** Utz Fischer, Qing Liu, and Gideon Dreyfuss, "The SMN-SIP1 Complex Has an Essential Role in Spliceosomal snRNP Biogenesis," *Cell* 90, 1023–29, 1997.
42 **it's not obvious why motor neurons:** Helena Chaytow, Yu-Ting Huang, Thomas H. Gillingwater, and Kiterie M. E. Faller, "The Role of Survival Motor Neuron Protein (SMN) in Protein Homeostasis," *Cellular and Molecular Life Sciences* 75, 3877–94, 2018.
42 **might be restored:** Luca Cartegni and Adrian R. Krainer, "Disruption of an SF2/ASF-Dependent Exonic Splicing Enhancer in SMN2 Causes Spinal Muscular Atrophy in the Absence of SMN1," *Nature Genetics* 30, 377–84, 2002.
43 **turned the concept into reality:** Yimin Hua, Kentaro Sahashi, Gene Hung, Frank Rigo, Marco A. Passini, C. Frank Bennett, and Adrian R. Krainer, "Antisense Correction of SMN2 Splicing in the CNS Rescues Necrosis in a Type III SMA Mouse Model," *Genes and Development* 24, 1634–44, 2010.
43 **"crawled all the way from the den":** Peter Tarr, "She's My Little Fighter," *Harbor Transcript* (Cold Spring Harbor Laboratory) 36, 4–7, 2016.
44 **generate the missing protein:** Leonela Amoasii, John C. W. Hildyard, Hui Li, Efrain Sanchez-Ortiz, Alex Mireault, Daniel Caballero, Rachel Harron, Thaleia-Rengina Stathopoulou, Claire Massey, John M. Shelton, Rhonda Bassel-Duby, Richard J. Piercy, and Eric N. Olson, "Gene Editing Restores Dystrophin Expression in a Canine Model of Duchenne Muscular Dystrophy," *Science* 362, 86–91, 2018.

CHAPTER 3: GOING IT ALONE

48 **detailing the action of enzymes:** J. B. S. Haldane, *Enzymes* (London: Longmans Green, 1930).
48 **enzymes consisted of:** David Blow, "So Do We Understand How Enzymes Work?," *Structure* 8, R77–R81, 2000.
48 **"All enzymes are proteins":** James B. Sumner, "The Chemical Nature of Enzymes," Nobel Lecture, December 12, 1946, https://www.nobelprize.org/uploads/2018/06/sumner-lecture.pdf.
50 **gene for its ribosomal RNA:** Joseph G. Gall, "Free Ribosomal RNA Genes in the Macronucleus of *Tetrahymena*," *Proceedings of the National Academy of Sciences USA* 71, 3078–81, 1974; Jan Engberg, Gunna Christiansen, and Vagn Leick, "Autonomous rDNA Molecules Containing Single Copies of the Ribosomal RNA Genes in the Macronucleus of *Tetrahymena pyriformis*," *Biochemical and Biophysical Research Communications* 59, 1356, 1974.
51 **about 400 base pairs:** Thomas R. Cech and Donald C. Rio, "Localization of Transcribed Regions on Extrachromosomal Ribosomal RNA Genes of *Tetrahymena thermophila* by R-loop Mapping," *Proceedings of the National Academy of Sciences USA* 76, 5051–55, 1979. A similar intron had been reported in a different species of *Tetrahymena*: Martha A. Wild and Joseph G. Gall, "An Intervening

Sequence in the Gene Coding for 25S Ribosomal RNA of *Tetrahymena pigmentosa*," *Cell* 16, 565–73, 1979.

53 **splicing of yeast mRNAs in a test tube:** Author's interview with John Abelson, San Francisco, California, March 25, 2022.

56 **tiny circles of RNA:** Paula J. Grabowski, Arthur J. Zaug and Thomas R. Cech, "The Intervening Sequence of the Ribosomal RNA Precursor Is Converted to a Circular RNA in Isolated Nuclei of *Tetrahymena*," *Cell* 23, 467–76, 1981.

57 **individual phosphorus and oxygen atoms:** Thomas R. Cech, Arthur J. Zaug, and Paula J. Grabowski, "In Vitro Splicing of the Ribosomal RNA Precursor of *Tetrahymena*: Involvement of a Guanosine Nucleotide in the Excision of the Intervening Sequence," *Cell* 27, 487–96, 1981.

60 **no protein enzyme involved:** Kelly Kruger, Paula J. Grabowski, Arthur J. Zaug, Julie Sands, Daniel E. Gottschling, and Thomas R. Cech, "Self-Splicing RNA: Autoexcision and Autocyclization of the Ribosomal RNA Intervening Sequence of *Tetrahymena*," *Cell* 31, 147–57, 1982.

60 **a fungus:** Gian Garriga and Alan M. Lambowitz, "RNA Splicing in *Neurospora* Mitochondria: Self-Splicing of a Mitochondrial Intron In Vitro," *Cell* 39, 631–41, 1984.

60 **baker's yeast:** Henk F. Tabak, G. Van der Horst, K. A. Osinga, and A. C. Arnberg, "Splicing of Large Ribosomal Precursor RNA and Processing of Intron RNA in Yeast Mitochondria," *Cell* 39, 623–29, 1984.

61 **even a bacterial virus:** Jonathan M. Gott, David A. Shub, and Marlene Belfort, "Multiple Self-Splicing Introns in Bacteriophage T4: Evidence from Autocatalytic GTP Labeling of RNA In Vitro," *Cell* 47, 81–87, 1986.

62 **remove the RNA from his preparations of the enzyme:** William H. McClain, Lien B. Lai, and Venkat Gopalan, "Trials, Travails and Triumphs: An Account of RNA Catalysis in RNase P," *Journal of Molecular Biology* 397, 627–46, 2010.

62 **essential for RNase P enzymatic activity:** Benjamin C. Stark, Ryszard Kole, E. J. Bowman, and Sidney Altman, "Ribonuclease P: An Enzyme with an Essential RNA Component," *Proceedings of the National Academy of Sciences USA* 75, 3717–21, 1978.

63 **need protein *and* RNA?:** Ryszard Kole and Sidney Altman, "Properties of Purified Ribonuclease P from *Escherichia coli*," *Biochemistry* 20, 1902–6, 1981.

63 **again give no activity:** Author's telephone interview with Cecilia Guerrier-Takada, Bethesda, Maryland, April 15, 2022.

63 **extra magnesium chloride:** Author's in-person interview with Norman Pace, Boulder, Colorado, April 15, 2022.

64 **precise trimming of the tRNA precursor:** Cecilia Guerrier-Takada, Kathleen Gardiner, Terry Marsh, Norman Pace, and Sidney Altman, "The RNA Moiety of Ribonuclease P Is the Catalytic Subunit of the Enzyme," *Cell* 35, 849–57, 1983.

64 **in his worldview, as well:** Author's in-person interview with Pace.

65 **their small size:** Anthony C. Forster and Robert H. Symons, "Self-Cleavage of Plus and Minus RNAs of a Virusoid and a Structural Model for the Active Sites,"

Cell 49, 211–20, 1987; Olke C. Uhlenbeck, "A Small Catalytic Oligoribonucleotide," *Nature* 328, 596–600, 1987; Jim Haseloff and Wayne L. Gerlach, "Simple RNA Enzymes with New and Highly Specific Endoribonuclease Activities," *Nature* 334, 585–51, 1988.

65 **"Girls can't do biochemistry"**: Christine Guthrie, "From the Ribosome to the Spliceosome and Back Again," *Journal of Biological Chemistry* 285, 1–12, 2010.

66 **unrelated to that of humans:** Guthrie, "From the Ribosome to the Spliceosome and Back Again," 3.

66 **U5 must be involved:** Bruce Patterson and Christine Guthrie, "An Essential Yeast snRNA with a U5-like Domain Is Required for Splicing," *Cell* 49, 613–24, 1987.

66 **soon more snRNAs were found:** Manuel Ares, Jr., "U2 RNA from Yeast Is Unexpectedly Large and Contains Homology to Vertebrate U4, U5 and U6 Small Nuclear RNAs," *Cell* 47, 49–59, 1986. This was the authentic yeast U2 RNA, but its proposed relationship to U4, U5, and U6 snRNAs did not turn out to be relevant. Also, in 1987 both the Guthrie lab and Mike Rosbash's lab at Brandeis found the elusive yeast U1.

66 **U1 bound to the left end of each intron:** Yuan Zhuang and Alan M. Weiner, "A Compensatory Base Change in U1 snRNA Suppresses a 5′ Splice Site Mutation," *Cell* 46, 827–35, 1986.

66 **U2 bound near the right end of each intron:** Roy Parker, Paul G. Siliciano, and Christine Guthrie, "Recognition of the TACTAAC Box During mRNA Splicing in Yeast Involves Base-Pairing with the U2-like snRNA," *Cell* 49, 229–39, 1987.

67 **the snRNA called U6:** Hiten D. Madhani and Christine Guthrie, "A Novel Base-Pairing Interaction Between U2 and U6 snRNAs Suggests a Mechanism for the Catalytic Activation of the Spliceosome," *Cell* 71, 803–17, 1992.

67 **"And he kind of got it!":** Author's Zoom interview with Christine Guthrie and John Abelson, San Francisco, California, March 25, 2022.

CHAPTER 4: THE SHAPE OF A SHAPESHIFTER

69 **"Form follows function":** Louis H. Sullivan, "The Tall Office Building Artistically Considered" (1896), in *Louis H. Sullivan, Kindergarten Chats and Other Writings*, ed. Isabella Athey (New York: George Wittenborn, 1979).

70 **as an encore:** Alexander Rich and J. D. Watson, "Some Relations Between DNA and RNA," *Proceedings of the National Academy of Sciences USA* 40, 759–64, 1954.

71 **diverse RNAs had a single, common structure:** Rich and Watson, "Some Relations Between DNA and RNA."

71 **purified ribosomes:** J. D. Watson, "Involvement of RNA in the Synthesis of Proteins," *Science* 140, 17–26, 1963.

71 **never been done for any type of RNA:** M. B. Hoagland, P. C. Zamecnik, and M. L. Stephenson, "Intermediate Reactions in Protein Biosynthesis," *Biochimica et Biophysica Acta* 24, 215–16, 1957; Mahlon B. Hoagland, Mary Louise Stephen-

son, Jesse F. Scott, Liselotte I. Hecht, and Paul C. Zamecnik, "A Soluble Ribonucleic Acid Intermediate in Protein Synthesis," *Journal of Biological Chemistry* 231, 241–57, 1958; Kikuo Ogata and Hiroyoshi Nohara, "The Possible Role of the Ribonucleic Acid (RNA) of the pH 5 Enzyme in Amino Acid Activation," *Biochimica et Biophysica Acta* 25, 659–60, 1957.

72 **mass of a raisin:** Robert W. Holley, "Alanine Transfer RNA," Nobel Lecture, December 12, 1968, https://www.nobelprize.org/uploads/2018/06/holley-lecture.pdf.

72 **amino acid alanine:** Holley, "Alanine Transfer RNA."

73 **to pair easily with the mRNA:** Holley, "Alanine Transfer RNA."

73 **Soon sequences of a dozen other types of tRNA:** Hans Georg Zachau, Dieter Dütting, and Horst Feldman, "The Structures of Two Serine Transfer Ribonucleic Acids," *Hoppe-Seyler's Zeitschrift für Physiologische Chemie* 347, 212–35, 1966; J. T. Madison, G. A. Everett, and H. K. Kung, "Nucleotide Sequence of a Yeast Tyrosine Transfer RNA," *Science* 153, 531–34, 1966; U. L. RajBhandary, S. H. Chang, A. Stuart, R. D. Faulkner, R. M. Hoskinson, and H. G. Khorana, "Studies on Polynucleotides, LXVIII. The Primary Structure of Yeast Phenylalanine Transfer RNA," *Proceedings of the National Academy of Sciences USA* 57, 751–58, 1967; Howard M. Goodman, John Abelson, Arthur Landy, S. Brenner, and J. D. Smith, "Amber Suppression: A Nucleotide Change in the Anticodon of a Tyrosine Transfer RNA," *Nature* 217, 1019–24, 1968; S. K. Dube, K. A. Marcker, B. F. C. Clark, and S. Cory, "Nucleotide Sequence of N-formyl-methionyl-transfer RNA," *Nature* 218, 232–33, 1968; S. Takemura, T. Mizutani, and M. Miyazaki, "The Primary Structure of Valine-I Transfer Ribonucleic Acid from *Torulopsis utilis*," *Journal of Biochemistry* 64, 277–78, 1968; M. Staehelin, H. Rogg, B. C. Baguley, T. Ginsberg, and W. Wehrli, "Structure of a Mammalian Serine tRNA," *Nature* 219, 1363–65, 1968.

75 **L-shaped molecule:** J. D. Robertus, Jane E. Ladner, J. T. Finch, Daniela Rhodes, R. S. Brown, B. F. C. Clark, and A. Klug, "Structure of Yeast Phenylalanine tRNA at 3 Å Resolution," *Nature* 250, 546–51, 1974; S. H. Kim, F. L. Suddath, G. J. Quigley, A. McPherson, J. L. Sussman, A. H. J. Wang, N. C. Seeman, and A. Rich, "Three-Dimensional Tertiary Structure of Yeast Phenylalanine Transfer RNA," *Science* 185, 435–40, 1974. The *Science* paper reports the structure completed by Sung-hou Kim after he had moved from MIT to Duke University.

76 **François proposed:** François Michel, Alain Jacquier, and Bernard Dujon, "Comparison of Fungal Mitochondrial Introns Reveals Extensive Homologies in RNA Secondary Structure," *Biochimie* 64, 867–81, 1982.

76 **his 2D model would also work:** Richard Waring and Wayne Davies published similar models at about the same time; see R. Wayne Davies, Richard B. Waring, John A. Ray, Terence Brown, and Claudio Scazzocchio, "Making Ends Meet: A Model for RNA Splicing in Fungal Mitochondria," *Nature* 300, 719–24, 1982; R. B. Waring, C. Scazzocchio, T. A. Brown, and R. W. Davies, "Close Relationship Between Certain Nuclear and Mitochondrial Introns," *Journal of Molecular Biology* 16, 595–605, 1983.

77 **a Dutch group confirmed this prediction:** Gerda van der Horst and Henk F. Tabak, "Self-Splicing of Yeast Mitochondrial Ribosomal and Messenger RNA Precursors," *Cell* 40, 759–66, 1985.

78 **curl up in his sleeping bag:** Author's interview with Eric Westhof, near Colmar, France, May 2, 2022.

78 **model of the *Tetrahymena* intron:** François Michel and Eric Westhof, "Modelling of the Three-Dimensional Architecture of Group I Catalytic Introns Based on Comparative Sequence Analysis," *Journal of Molecular Biology* 216, 585–610, 1990.

79 **"scissors" to cut out the intron:** François Michel, Maya Hanna, Rachel Green, David P. Bartel, and Jack W. Szostak, "The Guanosine Binding Site of the *Tetrahymena* Ribozyme," *Nature* 342, 391–95, 1989.

79 **hooked on science at an early age:** Sabin Russell, "Cracking the Code: Jennifer Doudna and Her Amazing Molecular Scissors," *California* (Cal Alumni Association), December 8, 2014, https://alumni.berkeley.edu/california-magazine/winter-2014-gender-assumptions/cracking-code-jennifer-doudna-and-her-amazing/.

80 **"P4-P6" that fit the bill:** Felicia L. Murphy and Thomas R. Cech, "An Independently Folding Domain of RNA Tertiary Structure Within the *Tetrahymena* Ribozyme," *Biochemistry* 32, 5291–5300, 1993; see also Felicia L. Murphy, Yuh-Hwa Wang, Jack D. Griffith, and Thomas R. Cech, "Coaxially Stacked RNA Helices in the Catalytic Center of the *Tetrahymena* Ribozyme," *Science* 265, 1709–12, 1994.

81 **a technical problem:** In order to solve the structure of a novel molecule of unknown shape, one needs not just one X-ray diffraction data set but also one of a "heavy-atom derivative" of the molecule. A heavy-atom derivative is the same molecule with some electron-dense atom residing at one or several fixed positions. Then, by comparing the diffraction pattern of the original molecule with the heavy-atom derivative, one can solve what is called the "crystallographic phase problem." This problem and its solution are detailed in Jennifer A. Doudna and Samuel H. Sternberg, *A Crack in Creation: Gene Editing and the Unthinkable Power to Control Evolution* (Boston: Houghton Mifflin Harcourt, 2017).

82 **specific sites in the RNA:** Jamie H. Cate and Jennifer A. Doudna, "Metal-Binding Sites in the Major Groove of a Large Ribozyme Domain," *Structure* 4, 1221–29, 1996.

83 **active as a biocatalyst:** Barbara L. Golden, Anne R. Gooding, Elaine R. Podell, and Thomas R. Cech, "A Preorganized Active Site in the Crystal Structure of the *Tetrahymena* Ribozyme," *Science* 282, 259–64, 1998; see also Feng Guo, Anne R. Gooding, and Thomas R. Cech, "Structure of the *Tetrahymena* Ribozyme: Base Triple Sandwich and Metal Ion at the Active Site," *Molecular Cell* 16, 351–62, 2004.

83 **functional RNA molecules were solved:** Adrian Ferré-D'Amaré, Kaihong Zhou, and Jennifer A. Doudna, "Crystal Structure of a Hepatitis Delta Virus Ribozyme," *Nature* 395, 567–74, 1998.

83 **RNA-only structures were reported:** Rhiju Das, "RNA Structure: A Renaissance Begins?," *Nature Methods* 18, 436, 2021. Despite its name, the RCSB Protein Data Bank (PDB; https://www.rcsb.org/) is also a repository for RNA sequences. Deposition of structures in the PDB is required when structures are published, so most structures solved by academics can be found there but only a small fraction of the structures (such as drug-protein complexes) solved in industrial labs.

84 **solutions to the RNA-folding problem:** Rhiju Das and Adrien Treuille had been postdoctoral fellows together in David Baker's lab at the University of Washington, where the crowdsourced protein-folding game called FoldIt had been invented. Now they each had their own faculty position, Das at Stanford and Treuille at Carnegie Mellon University in Pittsburgh, and the idea for eteRNA arose from a brainstorming session with Treuille's student Jeehyung Lee.

85 **100 eteRNA game-players as coauthors:** Jeehyung Lee, Wipapat Kladwang, Minjae Lee, Daniel Cantu, Martin Azizyan, Hanjoo Kim, Alex Limpaecher, Snehal Gaikwad, Sungroh Yoon, Adrien Treuille, Rhiju Das, and EteRNA Participants, "RNA Design Rules from a Massive Open Laboratory," *Proceedings of the National Academy of Sciences USA* 111, 2122–27, 2014.

86 **current computer programs:** Kathrin Leppek, Gun Woo Byeon, Wipapat Kladwang . . . and Rhiju Das, "Combinatorial Optimization of mRNA Structure, Stability and Translation for RNA-Based Therapeutics," *Nature Communications* 13, 1536, 2022.

86 **Now the Stanford researchers:** Rhiju Das, talk given at Nucleic Acids Chemistry and Biomedicine Symposium, University of Colorado, Boulder, September 17, 2022.

87 **and his Stanford colleague Ron Dror:** Raphael J. L. Townshend, Stephan Eismann, Andrew M. Watkins, Ramya Rangan, Masha Karelina, Rhiju Das, and Ron O. Dror, "Geometric Deep Learning of RNA Structure," *Science* 373, 1047–51, 2021.

88 **which he calls RNA-Puzzles:** Zhichao Miao, Ryszard W. Adamiak, Maciej Antczak . . . and Eric Westhof, "RNA-Puzzles Round IV: 3D Structure Predictions of Four Ribozymes and Two Aptamers," *RNA* 26, 982–95, 2020.

CHAPTER 5: THE MOTHERSHIP

92 **the hard work of protein synthesis:** Author's Zoom interview with Harry Noller, Santa Cruz, California, May 12, 2022.

92 **proteins organize themselves:** As one example, see R. A. Garrett and H. G. Wittmann, "Structure of Bacterial Ribosomes," *Advances in Protein Chemistry* 27, 277–347, 1973.

93 **dead in its tracks:** Author's Zoom interview with Noller.

93 **derail protein synthesis:** Harry F. Noller and Jonathan B. Chaires, "Functional Modification of 16S Ribosomal RNA by Kethoxal," *Proceedings of the National Academy of Sciences USA* 69, 3115–18, 1972.

NOTES

93 **key job of binding tRNAs:** Noller and Chaires, "Functional Modification of 16S Ribosomal RNA by Kethoxal."

93 **"Men occasionally stumble over the truth":** Quote Investigator, accessed September 2, 2023, https://quoteinvestigator.com/2012/05/26/stumble-over-truth/.

94 **worked for *all* the dozen organisms:** George E. Fox and Carl R. Woese, "5S RNA Secondary Structure," *Nature* 256, 505–7, 1975. Their proposed RNA structure differed from the ones that all the chemists had predicted. Fox and Woese correctly reasoned that the chemical approaches had purified the ribosomal RNA, removing it from its natural environment, and therefore perturbed its structure. The evolutionary evidence gleaned by comparing various sequences, on the other hand, pertained to the molecule in its natural habitat within the ribosome.

94 **take years:** Author's Zoom interview with Noller.

94 **"They didn't take us seriously," said Harry:** Author's Zoom interview with Noller.

95 **announced in 1980:** C. R. Woese, L. J. Magrum, R. Gupta, R. B. Siegel, D. A. Stahl, J. Kop, N. Crawford, J. Brosius, R. Gutell, J. J. Hogan, and H. F. Noller, "Secondary Structure Model for Bacterial 16S Ribosomal RNA: Phylogenetic, Enzymatic and Chemical Evidence," *Nucleic Acids Research* 8, 2275–93, 1980.

95 **the 2,904-nucleotide ribosomal RNA:** Harry F. Noller, JoAnn Kop, Virginia Wheaton, Jürgen Brosius, Robin R. Gutell, Alexei M. Kopylov, Ferdinand Dohme, Winship Herr, David A. Stahl, Ramesh Gupta, and Carl R. Woese, "Secondary Structure Model for 23S Ribosomal RNA," *Nucleic Acids Research* 9, 6167–89, 1981.

97 **southern Italian lizards:** C. Taddei, "Ribosome Arrangement During Oogenesis of *Lacerta sicula* Raf," *Experimental Cell Research* 70, 285–92, 1972; Ada E. Yonath, "Hibernating Bears, Antibiotics and the Evolving Ribosome," Nobel Lecture, December 8, 2009, https://www.nobelprize.org/uploads/2018/06/yonath_lecture.pdf. For a different view of ribosome crystals in hibernating bears, see Venki Ramakrishnan, *Gene Machine* (New York: Basic Books, 2018).

97 **crystallize in the laboratory:** Yonath, "Hibernating Bears, Antibiotics and the Evolving Ribosome."

97 **Yonath was able to grow crystals:** A. Yonath, J. Muessig, B. Tesche, S. Lorenz, V. A. Erdmann, and H. G. Wittmann, "Crystallization of the Large Ribosomal Subunit from *B. stearothermophilus*," *Biochemistry International* 1, 315–428, 1980.

97 **turned out to be winners:** A. Shevack, H. S. Gewitz, B. Hennemann, A. Yonath, and H. G. Wittmann, "Characterization and Crystallization of Ribosomal Particles from *Halobacterium marismortui*," *FEBS Letters* 184, 68–71, 1985.

98 **three postdoctoral fellows:** Nenad Ban came from Croatia via the University of California Riverside, Poul Nissen from Aarhus University in Denmark, and Jeff Hansen from the University of Colorado Boulder.

99 **the *Queen Mary* of biomolecular machines:** Thomas A. Steitz, "From the Structure and Function of the Ribosome to New Antibiotics," Nobel Lecture, December 8, 2009, https://www.nobelprize.org/uploads/2018/06/steitz_lecture.pdf.

99 **unimaginably sharp image:** N. Ban, P. Nissen, J. Hansen, P. B. Moore, and T. A.

Steitz, "The Complete Atomic Structure of the Large Ribosomal Subunit at 2.4 Å Resolution," *Science* 289, 905–20, 2000.
100 **a catalytic RNA machine:** Ban et al., "The Complete Atomic Structure of the Large Ribosomal Subunit." Also see Thomas R. Cech, "The Ribosome Is a Ribozyme," *Science* 289, 878–79, 2000.
100 **ribosome small subunit:** Brian T. Wimberly, Ditlev E. Brodersen, William M. Clemons, Jr., Robert J. Morgan-Warren, Andrew P. Carter, Clemens Vonrhein, Thomas Hartsch, and V. Ramakrishnan, "Structure of the 30S Ribosomal Subunit," *Nature* 407, 327–39, 2000.
101 **inhabited by tRNAs and mRNA:** Jamie H. Cate, Marat M. Yusupov, Gulnara Z. Yusupova, Thomas N. Earnest, and Harry F. Noller, "X-ray Crystal Structures of 70S Ribosome Functional Complexes," *Science* 285, 2095–104, 1999.
101 **complemented each other brilliantly:** V. Ramakrishnan, "Unraveling the Structure of the Ribosome," Nobel Lecture, December 8, 2009, https://www.nobelprize.org/uploads/2018/06/ramakrishnan_lecture.pdf.
102 **target bacterial ribosomes:** Steitz, "From the Structure and Function of the Ribosome to New Antibiotics."
104 **bound to its catalytic center:** Jeffrey L. Hanson, Peter B. Moore, and Thomas A. Steitz, "Structure of Five Antibiotics Bound at the Peptidyl Transferase Center of the Large Ribosomal Subunit," *Journal of Molecular Biology* 330, 1061–75, 2003.
104 **specific site on the large ribosomal subunit:** Jeffrey L. Hanson, T. Martin Schmeing, Peter B. Moore, and Thomas A. Steitz, "Structural Insights into Peptide Bond Formation," *Proceedings of the National Academy of Sciences USA* 99, 11670–75, 2002.
105 **"molecular constipation":** Steitz, "From the Structure and Function of the Ribosome to New Antibiotics."
105 **stuck to the RNA of the small ribosomal subunit:** Andrew P. Carter, William M. Clemons, Ditlev E. Brodersen, Robert J. Morgan-Warren, Brian T. Wimberly, and V. Ramakrishnan, "Functional Insights from the Structure of the 30S Ribosomal Subunit and Its Interactions with Antibiotics," *Nature* 407, 340–48, 2000; Ditlev E. Brodersen, William M. Clemons, Jr., Andrew P. Carter, Robert J. Morgan-Warren, Brian T. Wimberly, and V. Ramakrishnan, "The Structural Basis for the Action of the Antibiotics Tetracycline, Pactamycin, and Hygromycin B on the 30S Ribosomal Subunit," *Cell* 103, 1143–54, 2000.

CHAPTER 6: ORIGINS

110 **people trying to define it:** Frances Westall and André Brack, "The Importance of Water for Life," *Space Science Reviews* 214, 50, 2018.
112 **our key paper about the ribozyme:** Kelly Kruger, Paula J. Grabowski, Arthur J. Zaug, Julie Sands, Daniel E. Gottschling, and Thomas R. Cech, "Self-Splicing RNA: Autoexcision and Autocyclization of the Ribosomal RNA Intervening Sequence of *Tetrahymena*," *Cell* 31, 147–57, 1982.

113 **3.3 to 3.5 billion years ago:** J. William Schopf and Bonnie M. Packer, "Early Archean (3.3-Billion to 3.5-Billion-Year-Old) Microfossils from Warrawoona Group, Australia," *Science* 237, 70–73, 1987.

114 **an oft-cited 1968 paper:** Leslie E. Orgel, "Evolution of the Genetic Apparatus," *Journal of Molecular Biology* 38, 381–93, 1968.

114 **proteins and DNA came later:** Walter Gilbert, "The RNA World," *Nature* 319, 618, 1986.

117 **with an electric spark:** Stanley L. Miller and Harold C. Urey, "Organic Compound Synthesis on the Primitive Earth," *Science* 130, 245–51, 1959.

117 **can react to form nucleotides:** Matthew W. Powner, Beatrice Gerland, and John D. Sutherland, "Synthesis of Activated Pyrimidine Ribonucleotides in Prebiotically Plausible Conditions," *Nature* 459, 239–42, 2009.

117 **essentially in one pot:** Sidney Becker, Jonas Feldmann, Stefan Wiedemann, Hidenori Okamura, Christina Schneider, Katharina Iwan, Anthony Crisp, Martin Rossa, Tynchtyk Amatov, and Thomas Carell, "Unified Prebiotically Plausible Synthesis of Pyrimidine and Purine RNA Ribonucleotides," *Science* 366, 76–82, 2019.

118 **without any enzyme to power their assembly:** Tan Inoue and Leslie E. Orgel, "A Non-enzymatic RNA Polymerase Model," *Science* 219, 859–62, 1984; Gerald F. Joyce and Leslie E. Orgel, "Non-enzymatic Template-Directed Synthesis on RNA Random Copolymers: Poly(C,A) Templates," *Journal of Molecular Biology* 202, 677–81, 1988.

120 **the intron part of the ribozyme could catalyze the cutting and pasting:** Arthur J. Zaug and Thomas R. Cech, "The Intervening Sequence RNA of *Tetrahymena* Is an Enzyme," *Science* 231, 470–75, 1986; Michael D. Been and Thomas R. Cech, "RNA as an RNA Polymerase: Net Elongation of an RNA Primer Catalyzed by the *Tetrahymena* Ribozyme," *Science* 239, 1412–16, 1988.

121 **a bacteriophage ribozyme, SunY:** Jonatha Y. Gott, David A Shub, and Marlene Belfort, "Multiple Self-Splicing Introns in Bacteriophage T4: Evidence from Autocatalytic GTP Labeling of RNA In Vitro," *Cell* 47, 61–87, 1986.

122 **machine capable of self-replication:** Jennifer A. Doudna, Sandra Couture, and Jack W. Szostak, "A Multisubunit Ribozyme That Is the Catalyst of and the Template for Complementary Strand RNA Synthesis," *Science* 251, 1605–8, 1991.

122 **a separate RNA molecule:** Jennifer A. Doudna and Jack W. Szostak, "RNA-Catalysed Synthesis of Complementary-Strand RNA," *Nature* 339, 519–22, 1989.

123 **nucleic acid is randomly encapsulated:** Charles L. Apel, David W. Deamer, and Michael N. Mautner, "Self-Assembled Vesicles of Monocarboxylic Acids and Alcohols: Conditions for Stability and for the Encapsulation of Biopolymers," *Biochimica et Biophysica Acta* 1559, 1–9, 2002.

123 **nucleic acids can form longer chains:** Sheref S. Mansy, Jason P. Schrum, Mathangi Krishnamurthy, Sylvia Tobe, Douglas A. Treco, and Jack W. Szostak, "Template-Directed Synthesis of a Genetic Polymer Within a Model Protocell," *Nature* 454, 122–25, 2008.

CHAPTER 7: IS THE FOUNTAIN OF YOUTH A DEATH TRAP?

129 **multibillion-dollar antiaging industry:** Emily Stewart, "How the Anti-aging Industry Turns You Into a Customer for Life," *Vox*, July 28, 2022, https://www.vox.com/the-goods/2022/7/28/23219258/anti-aging-cream-expensive-scam.

132 **sequence repeated many times:** Elizabeth H. Blackburn and Joseph G. Gall, "A Tandemly Repeated Sequence at the Termini of the Extrachromosomal Ribosomal RNA Genes of *Tetrahymena*," *Journal of Molecular Biology* 120, 33–53, 1978.

134 **was now stable in yeast:** Jack W. Szostak and Elizabeth H. Blackburn, "Cloning Yeast Telomeres on Linear Plasmid Vectors," *Cell* 29, 245–55, 1982.

134 **yeast's own telomeric sequence:** Janis Shampay, Jack W. Szostak, and Elizabeth H. Blackburn, "DNA Sequences of Telomeres Maintained in Yeast," *Nature* 310, 154–57, 1984; see also Richard W. Walmsley, Clarence S. M. Chant, Bik-Kwoon Tye, and Thomas D. Petes, "Unusual DNA Sequences Associated with the Ends of Yeast Chromosomes," *Nature* 310, 157–60, 1984.

135 **took a chance on her:** "Carol W. Greider—Biographical," NobelPrize.org, accessed September 4, 2023, https://www.nobelprize.org/prizes/medicine/2009/greider/biographical/.

136 **direct evidence for the enzyme that would later be dubbed *telomerase*:** Carol W. Greider and Elizabeth H. Blackburn, "Identification of a Specific Telomere Terminal Transferase Activity in *Tetrahymena* Extracts," *Cell* 43, 405–13, 1985.

136 **something new to report every half hour:** Carol W. Greider, "Telomerase Discovery: The Excitement of Putting Together Pieces of the Puzzle," Nobel Lecture, December 7, 2009, https://www.nobelprize.org/uploads/2018/06/greider_lecture.pdf.

136 **activity appeared to require RNA:** Carol W. Greider and Elizabeth H. Blackburn, "The Telomere Terminal Transferase of *Tetrahymena* Is a Ribonucleoprotein Enzyme with Two Kinds of Primer Specificity," *Cell* 51, 887–89, 1987.

137 **TTGGGG's at *Tetrahymena* telomeres:** Carol W. Greider and Elizabeth H. Blackburn, "A Telomeric Sequence in the RNA of *Tetrahymena* Telomerase Required for Telomere Repeat Synthesis," *Nature* 337, 331–37, 1989.

138 **that composed human telomeres:** Junli Feng, Walter D. Funk, Sy-Shi Wang, Scott L. Weinrich, Ariel A. Avilion, Choy-Pik Chiu, Robert R. Adams, Edwin Chang, Richard C. Allsopp, Jinghua Yu, Siyuan Le, Michael D. West, Calvin B. Harley, William H. Andrews, Carol W. Greider, and Bryant Villeponteau, "The RNA Component of Human Telomerase," *Science* 269, 1236–41, 1995.

138 **basement laboratory:** Stephan S. Hall, *Merchants of Immortality* (Boston: Houghton Mifflin, 2003), 17; Leonard Hayflick, "My First Chemistry Kit" [video interview], WebofStories.com, accessed September 4, 2023, https://www.webofstories.com/play/leonard.hayflick/2.

138 **divide a limited number of times:** L. Hayflick and P. S. Moorhead, "The Serial Cultivation of Human Diploid Cell Strains," *Experimental Cell Research* 25, 585–621, 1961.

139 **50 base pairs per cell division:** Calvin B. Harley, A. Bruce Futcher, and Carol W.

Greider, "Telomeres Shorten During Ageing of Human Fibroblasts," *Nature* 345, 458–60, 1990.

139 **telomerase activity is a hallmark of all kinds of cancer:** N. W. Kim, M. A. Piatyszek, K. R. Prowse, C. B. Harley, M. D. West, P. L. Ho, G. M. Coviello, W. E. Wright, S. L. Weinrich, and J. W. Shay, "Specific Association of Human Telomerase Activity with Immortal Cells and Cancer," *Science* 266, 2011–14, 1994.

141 **one of Switzerland's premier RNA scientists:** The scientist was Walter Keller. See, for example, Joachim Lingner, Josef Kellermann, and Walter Keller, "Cloning and Expression of the Essential Gene for Poly(A) Polymerase from *S. cerevisiae*," *Nature* 354, 496–98, 1991.

142 **RNA subunit from *Euplotes*:** Joachim Lingner, Laura L. Hendrick, and Thomas R. Cech, "Telomerase RNAs of Different Ciliates Have a Common Secondary Structure and a Permuted Template," *Genes & Development* 8, 1984–98, 1994.

142 **telomerase from any organism:** Joachim Lingner and Thomas R. Cech, "Purification of Telomerase from *Euplotes aediculatus*: Requirement for a Primer 3'-Overhang," *Proceedings of the National Academy of Sciences USA* 93, 10712–17, 1996.

145 **our *Science* paper:** Joachim Lingner, Timothy R. Hughes, Andrej Shevchenko, Matthias Mann, Victoria Lundblad, and Thomas R. Cech, "Reverse Transcriptase Motifs in the Catalytic Subunit of Telomerase," *Science* 276, 561–67, 1997.

146 **Our paper describing the gene:** Toru M. Nakamura, Gregg B. Morin, Karen B. Chapman, Scott L. Weinrich, William H. Andrews, Joachim Lingner, Calvin B. Harley, and Thomas R. Cech, "Telomerase Catalytic Subunit Homologs from Fission Yeast and Human," *Science* 277, 955–59, 1997

146 **a fine story on the human TERT:** Matthew Meyerson, Christopher M. Counter, Elinor Ng Eaton, Leif W. Ellisen, Philipp Steiner, Stephanie Dickinson Caddle, Liuda Ziaugra, Roderick L. Beijersbergen, Michael J. Davidoff, Qingyun Liu, Silvia Bacchetti, Daniel A. Haber, and Robert A. Weinberg, "hEST2, the Putative Human Telomerase Catalytic Subunit Gene, Is Up-Regulated in Tumor Cells and During Immortalization," *Cell* 90, 785–95, 1997.

146 **active telomerase, both in the laboratory and in living cells:** Scott L. Weinrich, Ron Pruzan, Libin Ma, Michel Ouellette, Valerie M. Tesmer, Shawn E. Holt, Andrea G. Bodnar, Serge Lichtsteiner, Nam W. Kim, James B. Trager, Rebecca D. Taylor, Ruben Carlos, William H. Andrews, Woodring E. Wright, Jerry W. Shay, Calvin B. Harley, and Gregg B. Morin, "Reconstitution of Human Telomerase with the Template RNA Component hTR and the Catalytic Protein Subunit hTRT," *Nature Genetics* 17, 498–502, 1997. See also Kathleen Collins and Leena Gandhi, "The Reverse Transcriptase Component of the *Tetrahymena* telomerase Ribonucleoprotein Complex," *Proceedings of the National Academy of Sciences USA* 95, 8485–90, 1998; Tracy M. Bryan, Karen J. Goodrich, and Thomas R. Cech, "Telomerase RNA Bound by Protein Motifs Specific to Telomerase Reverse Transcriptase," *Molecular Cell* 6, 493–99, 2000; Gaël Cristofari and Joachim Lingner,

"Telomere Length Homeostasis Requires That Telomerase Levels Are Limiting," *EMBO Journal* 25, 565–574, 2006.
146 **no end in sight:** Andrea G. Bodnar, Michel Ouellette, Maria Frolkis, Shawn E. Holt, Choy-Pik Chiu, Gregg B. Morin, Calvin B. Harley, Jerry W. Shay, Serge Lichtsteiner, and Woodring E. Wright, "Extension of Life-Span by Introduction of Telomerase into Normal Human Cells," *Science* 279, 349–52, 1998.
148 **wrap around Earth three times:** Rebecca Skloot, *The Immortal Life of Henrietta Lacks* (New York: Crown, 2010), 2.
150 **misread it—a lot:** Author's virtual interview with Franklin Huang, University of California, San Francisco, December 8, 2022.
150 **many other cancers had also stumbled into this lucky-for-them trick:** Franklin W. Huang, Eran Hodis, Mary Jue Xu, Gregory V. Kryukov, Lynda Chin, and Levi A. Garraway, "Highly Recurrent TERT Promoter Mutations in Human Melanoma," *Science* 339, 957–59, 2013. See also Susanne Horn, Adina Figl, P. Sivaramakrishna Rachakonda, Christine Fischer, Antje Sucker, Andreas Gast, Stephanie Kadel, Iris Moll, Eduardo Nagore, Kari Hemminki, Dirk Schadendorf, and Rajiv Kumar, "TERT Promoter Mutations in Familial and Sporadic Melanoma," *Science* 339, 959–61, 2013.
150 **a TERT promoter mutation signals a more aggressive disease:** Matthias Simon, Ismail Hosen, Konstantinos Gousias, Sivaramakrishna Rachakonda, Barbara Heidenreich, Marco Gessi, Johannes Schramm, Kari Hemminki, Andreas Waha, and Rajiv Kumar, "*TERT* Promoter Mutations: A Novel Independent Prognostic Factor in Primary Glioblastomas," *Neuro-Oncology* 17, 45–52, 2015.

CHAPTER 8: AS THE WORM TURNS

153 **the worms are transparent:** Prof. Ding Xue and Dr. Joyita Bhadra, discussions and demonstration of nematode worm microinjection, Department of Molecular, Cellular and Developmental Biology, University of Colorado Boulder, November 17, 2022.
153 **the day's mission:** Andrew Z. Fire, "How Cells Respond to Genetic Change or Catching Up with Change in the Subway and in the Genome: A Bedtime Story." The Dr. H. P. Heineken Prize for Biochemistry and Biophysics (Amsterdam: Stichting Alfred Heineken Fondsen, 2004), 21.
153 **popular tool for manipulation of gene expression:** Sidney Pestka, "Antisense RNA—History and Perspective," *Annals of the New York Academy of Sciences* 660, 251–62, 1992.
154 **sense RNA also interrupted gene expression:** Su Guo and Kenneth J. Kemphues, "*par*-1, a Gene Required for Establishing Polarity in *C. elegans* Embryos, Encodes a Putative Ser/Thr Kinase That Is Asymmetrically Distributed," *Cell* 81, 611–20, 1995.
154 **"somewhat far-fetched hypothesis":** Andrew Z. Fire, "Gene Silencing by Double Stranded RNA," Nobel Lecture, December 8, 2006, https://www.nobelprize.org/

uploads/2018/06/fire_lecture.pdf; see also https://mcb.berkeley.edu/seminars/cdb2010symposium/fire_lecture.pdf.

155 **twitch like crazy:** Andrew Fire, SiQun Xu, Mary K. Montgomery, Steven A. Kostas, Samuel E. Driver, and Craig C. Mello, "Potent and Specific Genetic Interference by Double-Stranded RNA in *Caenorhabditis elegans*," *Nature* 391, 806–11, 1998.

156 ***C. elegans*:** Sydney Brenner, "The Genetics of *Caenorhabditis elegans*," *Genetics* 77, 71–94, 1974.

157 **David Hirsh:** "Craig C. Mello—Biographical," NobelPrize.org, accessed September 5, 2023, https://www.nobelprize.org/prizes/medicine/2006/mello/biographical/.

159 **extremely small nematode RNAs:** "Tiny RNAs That Regulate Gene Function," description and acceptance remarks for the 2008 Albert Lasker Basic Medical Research Award, accessed September 5, 2023, https://laskerfoundation.org/winners/tiny-rnas-that-regulate-gene-function/.

160 **by slicing them up:** Thomas Tuschl, Phillip D. Zamore, Ruth Lehmann, David P. Bartel, and Phillip A. Sharp, "Targeted mRNA Degradation by Double-Stranded RNA In Vitro," *Genes & Development* 13, 3191–7, 1999.

161 **dozens of previously overlooked microRNAs:** Mariana Lagos-Quintana, Reinhard Rauhut, Winfried Lendeckel, and Thomas Tuschl, "Identification of Novel Genes Coding for Small Expressed RNAs," *Science* 294, 853–58, 2001. See the related paper by Nelson C. Lau, Lee P. Lim, Earl G. Weinstein, and David P. Bartel, "An Abundant Class of Tiny RNAs with Probable Regulatory Roles in *Caenorhabditis elegans*," *Science* 294, 858–62, 2001.

161 **2 percent of the human genome:** Alexander R. Palazzo and Eugene V. Koonin, "Functional Long Non-coding RNAs Evolve from Junk Transcripts," *Cell* 183, 1151–61, 2020.

161 **and pregnancy:** Ramesh A. Shivdasani, "MicroRNAs: Regulators of Gene Expression and Cell Differentiation," *Blood* 108, 3646–53, 2006.

161 **inappropriate cell proliferation:** Lin He, Xingyue He, Lee P. Lim, Elisa de Stanchina, Zhenyu Xuan, Yu Liang, Wen Xue, Lars Zender, Jill Magnus, Dana Ridzon, Aimee L. Jackson, Peter S. Linsley, Caifu Chen, Scott W. Lowe, Michele A. Cleary, and Gregory J. Hannon, "A microRNA Component of the p53 Tumour Suppressor Network," *Nature* 447, 1130–34, 2007.

162 **the scientific groundwork for turning RNA into a therapeutic:** Sayda M. Elbashir, Jens Harborth, Winfried Lendeckel, Abdullah Yalcin, Klaus Weber, and Thomas Tuschl, "Duplexes of 21-Nucleotide RNAs Mediate RNA Interference in Cultured Mammalian Cells," *Nature* 411, 494–98, 2001. See also Sayda M. Elbashir, Winfried Lendeckel, and Thomas Tuschl, "RNA Interference Is Mediated by 21- and 22-Nucleotide RNAs," *Genes & Development* 15, 188–200, 2001.

163 **mutation in a single gene:** Jessica X. Chong, Kati J. Buckingham, Shalini N. Jhangiani . . . and Michael J. Bamshed, "The Genetic Basis of Mendelian Pheno-

types: Discoveries, Challenges, and Opportunities," *American Journal of Human Genetics* 97, 199–215, 2015.

163 **vitamin A, and other molecules:** Marcia Almeida Liz, Teresa Coelho, Vittorio Bellotti, Maria Isabel Fernandez-Arias, Pablo Mallaina, and Laura Obici, "A Narrative Review of the Role of Transthyretin in Health and Disease," *Neurology and Therapy* 9, 395–402, 2020.

164 **TTR protein from being produced:** David Adams, Ole B. Suhr, Peter J. Dyck, William J. Litchy, Raina G. Leahy, Jihong Chen, Jared Gollob, and Teresa Coelho, "Trial Design and Rationale for APOLLO, a Phase 3, Placebo-Controlled Study of Patisiran in Patients with Hereditary ATTR Amyloidosis with Polyneuropathy," *BMC Neurology* 17, 181, 2017.

164 **continued to deteriorate:** "Alnylam Reports Positive Topline Results from APOLLO-B Phase 3 Study of Patisiran in Patients with ATTR Amyloidosis with Cardiomyopathy" [press release], Alnylam, August 3, 2022, https://investors.alnylam.com/press-release?id=26851.

165 **cancer death rates in the United States decreased:** "An Update on Cancer Deaths in the United States," Centers for Disease Control and Prevention, last updated February 28, 2022, https://www.cdc.gov/cancer/dcpc/research/update-on-cancer-deaths/index.htm.

165 **Parkinson's:** Samuel A. Hasson, Lesley A. Kane, Koji Yamano, Chiu-Hui Huang, Danielle A. Sliter, Eugen Buehler, Chunxin Wang, Sabrina M. Heman-Ackah, Tara Hessa, Rajarshi Guha, Scott E. Martin, and Richard J. Youle, "High-Content Genome-Wide RNAi Screens Identify Regulators of Parkin Upstream of Mitophagy," *Nature* 504, 291–95, 2013.

165 **The death rate from Alzheimer's and Parkinson's:** *2023 Alzheimer's Disease Facts and Figures*, Alzheimer's Association, accessed September 5, 2023, https://www.alz.org/media/Documents/alzheimers-facts-and-figures.pdf.

165 **ALS is similarly on the rise:** Karissa C. Arthur, Andrea Calvo, T. Ryan Price, Joshua T. Geiger, Adriano Chio, and Bryan J. Traynor, "Projected Increase in Amyotrophic Lateral Sclerosis from 2015 to 2040," *Nature Communications* 7, 12408, 2016.

165 **technical name C9orf72:** Aaron R. Haeusler, Christopher J. Donnelly, and Jeffrey D. Rothstein, "The Expanding Biology of the C9orf72 Nucleotide Repeat Expansion in Neurodegenerative Disease," *Nature Reviews Neuroscience* 17, 383–95, 2016.

166 **on which neurons depend:** Haeusler et al., "The Expanding Biology of the C9orf72 Nucleotide Repeat Expansion."

167 **managed to silence amyloid precursor protein:** Kirk M. Brown, Jayaprakash K. Nair, Maja M. Janas . . . and Vasant Jadhav, "Expanding RNAi Therapeutics to Extrahepatic Tissues with Lipophilic Conjugates," *Nature Biotechnology* 40, 1500–1508, 2022.

CHAPTER 9: PRECISE PARASITES, SLOPPY COPIES

169 **the viral particles:** Wendell M. Stanley, "The Isolation and Properties of Crystalline Tobacco Mosaic Virus," Nobel Lecture, December 12, 1946, https://www.nobelprize.org/uploads/2018/06/stanley-lecture.pdf.
170 **crystallize the virus:** Wendell M. Stanley, "Isolation of a Crystalline Protein Possessing the Properties of Tobacco Mosaic Virus," *Science* 81, 644–45, 1935.
171 **stars in the known universe:** A. R. Mushegian, "Are There 10^{31} Virus Particles on Earth, or More, or Fewer?," *Journal of Bacteriology* 202, e00052-20, 2020.
171 **DNA is the "transforming principle":** Oswald T. Avery, Colin M. MacLeod, and Maclyn McCarty, "Studies on the Chemical Nature of the Substance Inducing Transformation of Pneumococcal Types: Induction of Transformation by a Desoxyribonucleic Acid Fraction Isolated from Pneumococcus Type III," *Journal of Experimental Medicine* 79, 137–58, 1944.
172 **cause infection with TMV:** A. Gierer and G. Schramm, "Infectivity of Ribonucleic Acid from Tobacco Mosaic Virus," *Nature* 177, 702–3, 1956.
172 **RNA is clearly the genetic material:** Heinz Fraenkel-Conrat, Beatrice A. Singer, and Robley C. Williams, "The Infectivity of Viral Nucleic Acid," *Biochimica et Biophysica Acta* 25, 87–96, 1957; Heinz Fraenkel-Conrat, Beatrice A. Singer, and Robley C. Williams, "The Nature of the Progeny of Virus Reconstituted from Protein and Nucleic Acid of Different Strains of Tobacco Mosaic Virus," in *Symposium on the Chemical Basis of Heredity*, ed. W. D. McElroy and B. Glass (Baltimore: Johns Hopkins University Press, 1957), 501–17.
175 **encoding 29 proteins:** Chongzhi Bai, Qiming Zhong, and George Fu Gao, "Overview of SARS-CoV-2 Genome-Encoded Proteins," *Science China Life Sciences* 65, 280–94, 2022.
176 **"as rapidly as possible?":** D. R. Mills, R. L. Peterson, and S. Spiegelman, "An Extracellular Darwinian Experiment with a Self-Duplicating Nucleic Acid Molecule," *Proceedings of the National Academy of Sciences USA* 58, 217–24, 1967.
177 **"little monster":** D. L. Kacian, D. R. Mills, F. R. Kramer, and S. Spiegelman, "A Replicating RNA Molecule Suitable for a Detailed Analysis of Extracellular Evolution and Replication," *Proceedings of the National Academy of Sciences USA* 69, 3038–42, 1972.
177 **replicating in the presence of ribonuclease:** Kacian et al., "A Replicating RNA Molecule."
177 **more infectious than earlier variants:** Lok Bahadur Shrestha, Charles Foster, William Rawlinson, Nicodemus Tedla, and Rowena A. Bull, "Evolution of the SARS-CoV-2 Variants BA.1 to BA.5: Implications for Immune Escape and Transmission," *Reviews in Medical Virology* 32, e2381, 2022.
178 **making therapeutic antibody treatments and vaccination less effective:** Masaud Shah and Hyun Goo Woo, "Omicron: A Heavily Mutated SARS-CoV-2 Variant Exhibits Stronger Binding to ACE2 and Potently Escapes Approved COVID-19 Therapeutic Antibodies," *Frontiers in Immunology* 12, 830527, 2022.

181 **600 progeny in about 8 hours:** Yinon M. Bar-On, Avi Flamholz, Rob Phillips, and Ron Milo, "SARS-CoV-2 (COVID-19) by the Numbers," *eLife* 9, e57309, 2020.

181 **laid flat by the viral infection:** Brandon Malone, Nadya Urakova, Eric J. Snijder, and Elizabeth A. Campbell, "Structures and Functions of Coronavirus Replication–Transcription Complexes and Their Relevance for SARS-CoV-2 Drug Design," *Nature Reviews Molecular Cell Biology* 23, 21–39, 2022.

CHAPTER 10: RNA VERSUS RNA

183 **"worthy of a visit by Picasso":** "History of Salk," Salk Institute, accessed September 5, 2023, https://www.salk.edu/about/history-of-salk.

184 **DNA vaccines:** Ellen F. Fynan, Shan Lu, and Harriet L. Robinson, "One Group's Historical Reflections on DNA Vaccine Development," *Human Gene Therapy* 29, 966–70, 2018.

184 **developed leukemia:** M. E. Gore, "Gene Therapy Can Cause Leukemia: No Shock, Mild Horror but a Probe," *Gene Therapy* 10, 4, 2003.

185 **saved millions of lives:** Eric C. Schneider, Arnav Shah, Pratha Sah, Seyed M. Moghadas, Thomas Vilches, and Alison P. Galvani, "The U.S. COVID-19 Vaccination Program at One Year: How Many Deaths and Hospitalizations Were Averted?," Issue Briefs, December 14, 2021, Commonwealth Fund, https://www.commonwealthfund.org/publications/issue-briefs/2021/dec/us-covid-19-vaccination-program-one-year-how-many-deaths-and.

188 **T7 RNA polymerase might be repurposed:** William T. McAllister, Claire Morris, Alan H. Rosenberg, and F. William Studier, "Utilization of Bacteriophage T7 Late Promoters in Recombinant Plasmids During Infection," *Journal of Molecular Biology* 153, 527–44, 1981.

188 **managed to isolate the gene of phage T7:** P. Davanloo, A. H. Rosenberg, J. J. Dunn, and F. W. Studier, "Cloning and Expression of the Gene for T7 RNA Polymerase," *Proceedings of the National Academy of Sciences USA* 81, 2035–39, 1984.

189 **performed in coffeehouses:** Author's interview with Philip Felgner, Irvine, California, October 20, 2022.

189 **a recipe for positively charged lipids:** P. L. Felgner, T. R. Gadek, M. Holm, R. Roman, H. W. Chan, M. Wenz, J. P. Northrop, G. M. Ringold, and M. Danielsen, "Lipofection: a Highly Efficient, Lipid-Mediated DNA-Transfection Procedure," *Proceedings of the National Academy of Sciences USA* 84, 7413–17, 1987.

189 **the year 2020:** Author's interview with Felgner.

190 **December 2020:** This initial approval was an emergency use authorization. Full FDA approval came in February 2022.

190 **translated into protein in its new home:** Robert W. Malone, Philip L. Felgner, and Inder M. Verma, "Cationic Liposome-Mediated RNA Transfection," *Proceedings of the National Academy of Sciences USA* 86, 1677–81, 1989.

190 **the muscle of mice:** Jon A. Wolff, Robert W. Malone, Phillip Williams, Wang

Chong, Gyula Acsadi, Agnes Jani, and Philip L. Felgner, "Direct Gene Transfer into Mouse Muscle In Vivo," *Science* 247, 1465–68, 1990.

191 **swept clean by cellular ribonuclease:** Fabrice Delaye, *The Medical Revolution of Messenger RNA* (Cold Spring Harbor, NY: Cold Spring Harbor Laboratory Press, 2023).

192 **infected by the flu virus:** Frédéric Martinon, Sivadasan Krishnan, Gerlinde Lenzen, Rémy Magné, Elisabeth Gomard, Jean-Gerard Guillet, Jean-Paul Lévy, and Pierre Meulien, "Induction of Virus-Specific Cytotoxic T Lymphocytes In Vivo by Liposome-Entrapped mRNA," *European Journal of Immunology* 23, 1719–22, 1993.

192 **bet on the future of DNA vaccines:** Delaye, *Medical Revolution of Messenger RNA*.

192 **when tested in animals:** H. Lv, S. Zhang, B. Wang, S. Cui, and J. Yan, "Toxicity of Cationic Lipids and Cationic Polymers in Gene Delivery," *Journal of Controlled Release* 114, 100–109, 2006.

193 **its RNA cargo is released into the cell's cytoplasm:** Pieter R. Cullis and Michael J. Hope, "Lipid Nanoparticle Systems for Enabling Gene Therapies," *Molecular Therapy* 25, 1467–75, 2017.

194 **distinguish it from normal human RNAs:** Chelsea M. Hull and Philip C. Bevilacqua, "Discriminating Self and Non-self by RNA: Roles for RNA Structure, Misfolding, and Modification in Regulating the Innate Immune Sensor PKR," *Accounts of Chemical Research* 49, 1242–49, 2016.

194 **nasty inflammatory response:** Katalin Karikó, Houping Ni, John Capodici, M. Lamphier, and Drew Weissman, "mRNA Is an Endogenous Ligand for Toll-Like Receptor 3," *Journal of Biological Chemistry* 279, 12542–50, 2004.

194 **teddy bear:** David Crow, "How mRNA Became a Vaccine Game-Changer," *FT Magazine*, May 13, 2021, https://www.ft.com/content/b2978026-4bc2-439c-a561-a1972eeba940.

194 **"always: no, no, no":** Damian Garde and Jonathan Saltzman, "The Story of mRNA: How a Once-Dismissed Idea Became a Leading Technology in the Covid Vaccine Race," STAT, November 10, 2020.

195 **matching interests:** Lasker-DeBakey Clinical Medical Research Award interview with Katalin Karikó, 2021.

195 **more reserved and methodical:** L'Oreal Award interview with Katalin Karikó, 2022.

195 **largely ignored by the innate immune system:** Katalin Karikó, Michael Buckstein, Houping Ni, and Drew Weissman, "Suppression of RNA Recognition by Toll-Like Receptors: The Impact of Nucleoside Modification and the Evolutionary Origin of RNA," *Immunity* 23, 165–75, 2005.

195 **more stable than the natural version:** Katalin Karikó, Hiromi Muramatsu, Frank A. Welsh, Janos Ludwig, Hiroki Kato, Shizuo Akira, and Drew Weissman, "Incorporation of Pseudouridine into mRNA Yields Superior Nonimmunogenic

Vector with Increased Translational Capacity and Biological Stability," *Molecular Therapy* 16, 1833–40, 2008.

195 **enhanced by pseudoU:** Others later found that the addition of a methyl group (a carbon atom and three hydrogens) to pseudoU further enhanced the mRNA's performance. See Oliwia Andries, Séan McCafferty, Stefaan C. De Smedt, Ron Weiss, Niek N. Sanders, and Tasuku Kitada, "N1-methylpseudouridine-Incorporated mRNA Outperforms Pseudouridine-Incorporated mRNA by Providing Enhanced Protein Expression and Reduced Immunogenicity in Mammalian Cell Lines and Mice," *Journal of Controlled Release* 217, 337–44, 2015.

195 **companies that licensed Karikó and Weissman's technology:** Alex Gardner, "Penn mRNA Scientists Drew Weissman and Katalin Karikó Receive 2021 Lasker Award, America's Top Biomedical Research Prize," *Penn Medicine News*, September 24, 2021.

196 **open-access website:** "Novel 2019 Coronavirus Genome," Virological.org, January 10, 2020, https://virological.org/t/novel-2019-coronavirus-genome/319.

196 **more of a threat than SARS or MERS:** Author's telephone interview with Dr. Melissa Moore, Moderna, Inc., June 15, 2022.

196 **no travel restrictions to contain the outbreak:** Özlem Türeci and Ugur Sahin, "Racing for a SARS-CoV-2 Vaccine," *EMBO Molecular Medicine* 13, e15145, 2021.

197 **proven technologies:** Damian Garde, "Covid-19 Drugs & Vaccines Tracker," STAT, accessed September 5, 2023, https://www.statnews.com/2020/04/27/drugs-vaccines-tracker/.

197 **Oxford-AstraZeneca DNA vaccine:** Jonathan Corum and Carl Zimmer, "How the Oxford-AstraZeneca Vaccine Works," *New York Times*, May 7, 2021, https://www.nytimes.com/interactive/2020/health/oxford-astrazeneca-covid-19-vaccine.html.

198 **The 90 protein spikes:** Yinon M. Bar-On, Avi Flamholz, Rob Phillips, and Ron Milo, "SARS-CoV-2 (Covid-19) by the Numbers," *eLife* 9, e57309, 2020.

198 **inflexible amino acids called prolines:** Jesper Pallesen, Nianshuang Wang, Kizzmekia S. Corbett, Daniel Wrapp, Robert N. Kirchdoerfer, Hannah L. Turner, Christopher A. Cottrell, Michelle M. Becker, Lingshu Wang, Wei Shi, Wing-Pui Kong, Erica L. Andres, Arminja N. Kettenbach, Mark R. Denison, James D. Chappell, Barney S. Graham, Andrew B. Ward, and Jason S. McLellan, "Immunogenicity and Structures of a Rationally Designed Prefusion MERS CoV Spike Antigen," *Proceedings of the National Academy of Sciences USA* 114, E7348–E7357, 2017.

199 **20 mRNA sequences to test:** Türeci and Sahin, "Racing for a SARS-CoV-2 Vaccine."

199 **and the world:** Garde and Saltzman, "Story of mRNA."

199 **shouts of triumph:** This anecdote was shared with the author by someone who attended the board meeting and prefers to remain anonymous.

199 **measles vaccine (97 percent effective):** "Past Seasons' Vaccine Effectiveness Estimates," Centers for Disease Control and Prevention, accessed September 5,

2023, https://www.cdc.gov/flu/vaccines-work/past-seasons-estimates.html. The flu vaccine data are from 2004–2022.

199 **similarly effective vaccines:** The slightly superior performance of the Moderna vaccine in some trials is probably due to the higher dose that was chosen to be administered, so it is not considered to be a distinguishing feature. See E. J. Rubin and D. L. Longo, "Covid-19 mRNA Vaccines—Six of One, Half a Dozen of the Other," *New England Journal of Medicine* 386, 183–85, 2022.

199 **"A shot to save the world":** Gregory Zuckerman, *A Shot to Save the World: The Inside Story of the Life-or-Death Race for a Covid-19 Vaccine* (New York: Penguin, 2021).

200 **"for vaccines against other infectious diseases":** "Prize Announcement," NobelPrize.org, accessed October 16, 2023, https://www.nobelprize.org/prizes/medicine/2023/prize-announcement/.

201 **once lost to cervical cancer:** "HPV and Cancer," National Cancer Institute, last updated April 4, 2023, https://www.cancer.gov/about-cancer/causes-prevention/risk/infectious-agents/hpv-and-cancer. The author discloses that he is a former member of the board of directors of Merck, Inc., and owns MRK stock.

202 **the lung cancer was dramatically curtailed:** David Boczkowski, Smita K. Nair, David Snyder, and Eli Gilboa, "Dendritic Cells Pulsed with RNA Are Potent Antigen-Presenting Cells In Vitro and In Vivo," *Journal of Experimental Medicine* 184, 465–72, 1996.

202 **no approved therapeutic yet:** Eli Gilboa, David Boczkowski, and Smita K. Nair, "The Quest for mRNA Vaccines," *Nucleic Acid Therapeutics* 32, 449–56, 2022.

202 **reenergized their mRNA cancer vaccine programs:** Ian Sample, "Vaccines to Treat Cancer Possible by 2030, Say BioNTech Founders," *The Guardian*, October 16, 2022; Julie Steenhuysen and Michael Erman, "Positive Moderna, Merck Cancer Vaccine Data Advances mRNA Promise, Shares Rise," Reuters, December 13, 2022.

202 **an mRNA vaccine aimed at melanoma:** Gina Vitale, "Moderna/Merck Cancer Vaccine Shows Promise in Trials," *Chemical & Engineering News*, December 20, 2022.

CHAPTER 11: RUNNING WITH SCISSORS

205 **recent books have touted:** Jennifer A. Doudna and Samuel H. Sternberg, *A Crack in Creation: Gene Editing and the Unthinkable Power to Control Evolution* (Boston: Houghton Mifflin Harcourt, 2017); Walter Isaacson, *The Code Breaker: Jennifer Doudna, Gene Editing, and the Future of the Human Race* (New York: Simon & Schuster, 2021); Kevin Davies, *Editing Humanity: The CRISPR Revolution and the New Era of Genome Editing* (New York: Pegasus Books, 2020).

206 **Doudna and Charpentier's discovery was announced:** Martin Jinek, Krzysztof Chylinski, Ines Fonfara, Michael Hauer, Jennifer A. Doudna, and Emmanuelle Charpentier, "A Programmable Dual-RNA–Guided DNA Endonuclease in Adaptive Bacterial Immunity," *Science* 337, 816–21, 2012.

NOTES

206 **Feng Zhang at MIT, George Church at Harvard, and Jennifer's own lab at Berkeley:** Le Cong, F. Ann Ran, David Cox, Shuailiang Lin, Robert Barretto, Naomi Habib, Patrick D. Hsu, Xuebing Wu, Wenyan Jiang, Luciano A. Marraffini, and Feng Zhang, "Multiplex Genome Engineering Using CRISPR/Cas Systems," *Science* 339, 819–23, 2013; Prashant Mali, Luhan Yang, Kevin M. Esvelt, John Aach, Marc Guell, James E. DiCarlo, Julie E. Norville, and George M. Church, "RNA-Guided Human Genome Engineering via Cas9," *Science* 339, 823–27, 2013; Martin Jinek, Alexandra East, Aaron Cheng, Steven Lin, Enbo Ma, and Jennifer Doudna, "RNA-Programmed Genome Editing in Human Cells," *eLife* 2, e00471, 2013.

207 **7,000 research laboratories worldwide:** Caroline M. LaManna and Rodolphe Barrangou, "Enabling the Rise of a CRISPR World," *CRISPR Journal* 1, 205–8, 2018. The number 7,000 is based on the number of laboratories receiving CRISPR plasmids from the Addgene distribution center from 2013 to 2018 and assumes a constant growth from 2013 to 2023. Because Addgene is only one source of CRISPR plasmids, this may substantially underestimate the number of laboratories using the technology.

208 **new phage sequences into the array:** Julie Grainy, Sandra Garrett, Brenton R. Graveley, and Michael P. Terns, "CRISPR Repeat Sequences and Relative Spacing Specify DNA Integration by *Pyrococcus furiosus* Cas1 and Cas2," *Nucleic Acids Research* 47, 7518–31, 2019.

208 **obliterate attacking phages:** Rodolphe Barrangou, Christophe Fremaux, Hélène Deveau, Melissa Richards, Patrick Boyaval, Sylvain Moineau, Dennis A. Romero, and Philippe Horvath, "CRISPR Provides Acquired Resistance Against Viruses in Prokaryotes," *Science* 315, 1709–12, 2007.

208 **destroy it:** Josiane E. Garneau, Marie-Ève Dupuis, Manuela Villion, Dennis A. Romero, Rodolphe Barrangou, Patrick Boyaval, Christophe Fremaux, Philippe Horvath, Alfonso H. Magadán, and Sylvain Moineau, "The CRISPR/Cas Bacterial Immune System Cleaves Bacteriophage and Plasmid DNA," *Nature* 468, 67–71, 2010.

208 **directed the antiviral defense:** Stan J. J. Brouns, Matthijs M. Jore, Magnus Lundgren, Edze R. Westra, Rik J. H. Slijkhuis, Ambrosius P. L. Snijders, Mark J. Dickman, Kira S. Makarova, Eugene V. Koonin, and John van der Oost, "Small CRISPR RNAs Guide Antiviral Defense in Prokaryotes," *Science* 321, 960–64, 2008.

208 **a leader in this nascent field:** Rachel E. Haurwitz, Martin Jinek, Blake Wiedenheft, Kaihong Zhou, and Jennifer A. Doudna, "Sequence- and Structure-Specific RNA Processing by a CRISPR Endonuclease," *Science* 329, 1355–58, 2010.

209 **conference in Puerto Rico:** Doudna and Sternberg, *Crack in Creation*, 70.

209 **Jennifer found Emmanuelle to be:** Doudna and Sternberg, *Crack in Creation*, 71.

210 **Czech-Polish border:** Doudna and Sternberg, *Crack in Creation*, 78.

210 **the specificity was exquisite:** Jinek et al., "A Programmable Dual-RNA–Guided DNA Endonuclease in Adaptive Bacterial Immunity."

211 **perfect match:** Although perfectly matched sequences are cleaved, the system will

tolerate some mismatches in the guide RNA–DNA pairing. See Patrick D. Hsu, David A. Scott, Joshua A. Weinstein, F. Ann Ran, Silvana Konermann, Vineeta Agarwala, Yinqing Li, Eli J. Fine, Xuebing Wu, Ophir Shalem, Thomas J. Cradick, Luciano A. Marraffini, Gang Bao, and Feng Zhang, "DNA Targeting Specificity of RNA-Guided Cas9 Nucleases," *Nature Biotechnology* 31, 827–32, 2013.

212 **publish their discovery:** Jinek et al., "A Programmable Dual-RNA–Guided DNA Endonuclease in Adaptive Bacterial Immunity."

212 **That's when the gene editing happens:** Author interview with Dana Carroll, Distinguished Professor, University of Utah, in Berkeley, California, January 13, 2023. See also Dana Carroll, "A CRISPR Approach to Gene Targeting," *Molecular Therapy* 20, 1658–60, 2012.

217 **hooked up known gene repressors and activators to Dead Cas9:** L. S. Qi, M. H. Larson, L. A. Gilbert, J. A. Doudna, J. S. Weissman, A. P. Arkin, and Wendell A. Lim, "Repurposing CRISPR as an RNA-Guided Platform for Sequence-Specific Control of Gene Expression," *Cell* 152, 1173–83, 2013.

217 **other Dead Cas9 tools:** Luke A. Gilbert, Matthew H. Larson, Leonardo Morsut, Zairan Liu, Gloria A. Brar, Sandra E. Torres, Noam Stern-Ginossar, Onn Brandman, Evan H. Whitehead, Jennifer A. Doudna, Wendell A. Lim, Jonathan S. Weissman, and Lei S. Qi, "CRISPR-Mediated Modular RNA-Guided Regulation of Transcription in Eukaryotes," *Cell* 154, 442–51, 2013; David Bikard, Wenyan Jiang, Poulami Samai, Ann Hochschild, Feng Zhang, and Luciano A. Marraffini, "Programmable Repression and Activation of Bacterial Gene Expression Using an Engineered CRISPR-Cas System," *Nucleic Acids Research* 41, 7429–37, 2013; L. A. Gilbert, M. A. Horlbeck, B. Adamson, J. E. Villalta, Y. Chen, E. H. Whitehead, C. Guimaraes, B. Panning, H. L. Ploegh, M. C. Bassik, L. S. Qi, M. Kampmann, and J. S. Weissman, "Genome-Scale CRISPR-Mediated Control of Gene Repression and Activation," *Cell* 159, 647–61, 2014.

219 **Cas12a:** Bijoya Paul and Guillermo Montoya, "CRISPR-Cas12a: Functional Overview and Applications," *Biomedical Journal* 43, 8–17, 2020.

219 **pinned down at the molecular level was sickle cell disease:** Linus Pauling, Harvey A. Itano, S. J. Singer, and Ibert C. Wells, "Sickle Cell Anemia: A Molecular Disease," *Science* 110, 543–48, 1949; Vernon M. Ingram, "Gene Mutations in Human Haemoglobin: The Chemical Difference Between Normal and Sickle Cell Haemoglobin," *Nature* 180, 326–28, 1957.

220 **companies have a sickle cell program:** The companies include Intellia, Beam Therapeutics, Editas Medicine, and CRISPR Therapeutics, which works with the pharma company Vertex.

220 **David Liu's base-editing technology:** Gregory A. Newby, Jonathan S. Yen, Kaitly J. Woodard . . . and David R. Liu, "Base Editing of Haematopoietic Stem Cells Rescues Sickle Cell Disease in Mice," *Nature* 595, 295–302, 2021.

220 **healthy competition:** Hope Henderson, "CRISPR Clinical Trials: A 2023 Update," Innovative Genomics, March 17, 2023, https://innovativegenomics.org/news/crispr-clinical-trials-2023/.

NOTES

221 **became the first sickle cell patient:** Rob Stein, "First Sickle Cell Patient Treated with CRISPR Gene-Editing Still Thriving," NPR, December 31, 2021, https://www.npr.org/sections/health-shots/2021/12/31/1067400512/first-sickle-cell-patient-treated-with-crispr-gene-editing-still-thriving. This particular CRISPR technology was developed by CRISPR Therapeutics and Vertex; see Haydar Frangoul, David Altshuler, M. Domenica Cappellini . . . and Selim Corbacioglu, "CRISPR-Cas9 Gene Editing for Sickle Cell Disease and β-Thalassemia," *New England Journal of Medicine* 384, 252–60, 2021.

221 **drug called Exa-cel:** Cormac Sheridan, "The World's First CRISPR Therapy Is Approved: Who Will Receive It? *Nature Biotechnology*, November 21, 2023. https://doi.org/10.1038/d41587-023-00016-6. Unlike some other CRISPR therapeutics in development, Exa-cel does not correct the mutation in the beta-globin gene. Instead, it inactivates a gene responsible for repressing the production of fetal beta-globin, thereby allowing the fetal protein to be expressed and substitute for the mutant adult beta-globin; see also Adam Zamecnik, "CRISPR Gene Therapies: Is 2023 a Milestone Year in the Making?," *Pharmaceutical Technology*, January 3, 2023.

222 **moved "too quickly":** Hannah Devlin, "Scientist Who Edited Babies' Gene Says He Acted 'Too Quickly,' " *The Guardian*, February 4, 2023.

222 **restricted to somatic cells:** David Baltimore, Paul Berg, Michael Botchan, Dana Carroll, R. Alta Charo, George Church, Jacob E. Corn, George Q. Daley, Jennifer A. Doudna, Marsha Fenner, Henry T. Greely, Martin Jinek, G. Steven Martin, Edward Penhoet, Jennifer Puck, Samuel H. Sternberg, Jonathan S. Weissman, and Keith R. Yamamoto, "A Prudent Path Forward for Genomic Engineering and Germline Gene Modification," *Science* 348, 36–38, 2015.

223 **Five gene therapies:** Asher Mullard, "FDA Approves First Haemophilia B Gene Therapy," *Nature Reviews Drug Discovery* 22, 7, 2023.

223 **human and nonhuman applications:** Baltimore et al., "A Prudent Path Forward."

224 **CRISPR machinery has the potential to do just that:** Andrew Hammond, Roberto Galizi, Kyros Kyrou, Alekos Simoni, Carla Siniscalchi, Dimitris Katsanos, Matthew Gribble, Dean Baker, Eric Marois, Steven Russell, Austin Burt, Nikolai Windbichler, Andrea Crisanti, and Tony Nolan, "A CRISPR-Cas9 Gene Drive System Targeting Female Reproduction in the Malaria Mosquito Vector *Anopheles gambiae*," *Nature Biotechnology* 34, 78–83, 2016; Rebecca Roberts and Brittany Enzmann, "CRISPR Gene Drives: Eradicating Malaria, Controlling Pests, and More," Synthego, August 9, 2022, https://www.synthego.com/blog/gene-drive-crispr.

225 **"silver bullet or a global conservation threat?":** Bruce L. Webber, S. Raghu, and Owain R. Edwards, "Opinion: Is CRISPR-Based Gene Drive a Biocontrol Silver Bullet or a Global Conservation Threat?," *Proceedings of the National Academy of Sciences USA* 112, 10565–67, 2015.

225 **other species of mosquitos and other insects:** C. M. Collins, J. A. S. Bonds, M. M. Quinlan, and J. D. Mumford, "Effects of the Removal or Reduction in Density of

the Malaria Mosquito, *Anopheles gambiae s.l.*," on Interacting Predators and Competitors in Local Ecosystems," *Medical and Veterinary Entomology* 33, 1–15, 2019.

226 **paid farmers to plant kudzu:** Nancy J. Loewenstein, Stephen F. Enloe, John W. Everest, James H. Miller, Donald M. Ball, and Michael G. Patterson, "History and Use of Kudzu in the Southeastern United States," *Forestry & Wildlife*, Alabama Cooperative Extension System, March 8, 2022, https://www.aces.edu/blog/topics/forestry-wildlife/the-history-and-use-of-kudzu-in-the-southeastern-united-states/.

227 **before CRISPR gene drive is implemented:** National Academies of Sciences, Engineering, and Medicine, *Gene Drives on the Horizon: Advancing Science, Navigating Uncertainty, and Aligning Research with Public Values* (Washington, DC: National Academies Press, 2016).

227 **the hottest years on record:** Henry Fountain and Mira Rojanasakul, "The Last 8 Years Were the Hottest on Record," *New York Times*, January 10, 2023, https://www.nytimes.com/interactive/2023/climate/earth-hottest-years.html.

228 **altering the coral genome:** Phillip A. Cleves, Marie E. Strader, Line K. Bay, John R. Pringle, and Mikhail V. Matz, "CRISPR/Cas9-Mediated Genome Editing in a Reef-Building Coral," *Proceedings of the National Academy of Sciences USA* 115, 5235–40, 2018.

228 **hydroelectric, solar, wind, and nuclear:** "U.S. Renewable Energy Factsheet," publication no. CSS03-12, Center for Sustainable Systems, University of Michigan, 2022.

228 **cancels out the benefit of its carbon fixation:** James Conca, "It's Final: Corn Ethanol Is of No Use," *Forbes*, April 20, 2014.

228 **freshwater required by corn:** Conca, "It's Final."

228 **production of this biofuel:** Sudarshan Singh Lakhawat, Naveen Malik, Vikram Kumar, Sunil Kumar, and Pushpender Kumar Sharma, "Implications of CRISPR-Cas9 in Developing Next Generation Biofuel: A Mini-review," *Current Protein and Peptide Science* 23, 574–84, 2022.

229 **divert these bacteria from releasing methane:** L. Val Giddings, Robert Rozansky, and David M. Hart, "Gene Editing for the Climate: Biological Solutions for Curbing Greenhouse Emissions," Information Technology & Innovation Foundation, September 2020, https://www2.itif.org/2020-gene-edited-climate-solutions.pdf.

229 **increase the carbon-storing capacity of the root systems:** "Supercharging Plants and Soils to Remove Carbon from the Atmosphere" [press release], Innovative Genomics Institute, June 14, 2022, https://innovativegenomics.org/news/crispr-carbon-removal/.

EPILOGUE: THE FUTURE OF RNA

231 **27 percent is dark matter, and 68 percent is dark energy:** Daniel Clery, "Into the Dark," *Science* 380, 1212–15, 2023.

232 **several hundred thousand:** Michael T. Y. Lam, Wenbo Li, Michael G. Rosenfeld, and Christopher K. Glass, "Enhancer RNAs and Regulated Transcriptional Programs," *Trends in Biochemical Sciences* 39, 170–82, 2014.

232 **prevents mice from fending off bacterial infections:** Jordan P. Lewandowski, James C. Lee, Taeyoung Hwang, Hongjae Sunwoo, Jill M. Goldstein, Abigail F. Groff, Nydia P. Chang, William Mallard, Adam Williams, Jorge Henao-Meija, Richard A. Flavell, Jeannie T. Lee, Chiara Gerhardinger, Amy J. Wagers, and John L. Rinn, "The *Firre* Locus Produces a *trans*-Acting RNA Molecule That Functions in Hematopoiesis," *Nature Communications* 10, 5137, 2019.

232 **lncRNA, called Tug1:** Jordan P. Lewandowski, Gabrijela Dumbović, Audrey R. Watson, Taeyoung Hwang, Emily Jacobs-Palmer, Nydia Chang, Christian Much, Kyle M. Turner, Christopher Kirby, Nimrod D. Rubinstein, Abigail F. Groff, Steve C. Liapis, Chiara Gerhardinger, Assaf Bester, Pier Paolo Pandolfi, John G. Clohessy, Hopi E. Hoekstra, Martin Sauvageau, and John L. Rinn, "The Tug1 lncRNA Locus Is Essential for Male Fertility," *Genome Biology* 21, 237, 2020.

234 **Larry Gold:** Craig Tuerk and Larry Gold, "Systematic Evolution of Ligands by Exponential Enrichment: RNA Ligands to Bacteriophage T4 DNA Polymerase," *Science* 249, 505–10, 1990.

234 **Jack Szostak:** Andrew D. Ellington and Jack W. Szostak, "In Vitro Selection of RNA Molecules That Bind Specific Ligands," *Nature* 346, 818–22, 1990.

235 **a company, SomaLogic:** The author discloses that at the time of publication of this book, he was on the Scientific Advisory Board of SomaLogic.

235 **advance warning of the progression of heart disease and various cancers:** Marie Cuvelliez, Vincent Vandewalle, Maxime Brunin, Olivia Beseme, Audrey Hulot, Pascal de Groote, Philippe Amouyel, Christophe Bauters, Guillemette Marot, and Florence Pinet, "Circulating Proteomic Signature of Early Death in Heart Failure Patients with Reduced Ejection Fraction," *Scientific Reports* 9, 19202, 2019; Anna Egerstedt, John Berntsson, Maya Landenhed Smith, Olof Gidlöf, Roland Nilsson, Mark Benson, Quinn S. Wells, Selvi Celik, Carl Lejonberg, Laurie Farrell, Sumita Sinha, Dongxiao Shen, Jakob Lundgren, Göran Rådegran, Debby Ngo, Gunnar Engström, Qiong Yang, Thomas J. Wang, Robert E. Gerszten, and J. Gustav Smith, "Profiling of the Plasma Proteome Across Different Stages of Human Heart Failure," *Nature Communications* 10, 5830, 2019.

235 **being developed as biosensors:** Frieder W. Scheller, Ulla Wollenberger, Axel Warsinke, and Fred Lisdat, "Research and Development in Biosensors," *Current Opinion in Biotechnology* 12, 35–40, 2001; Erin M. McConnell, Julie Nguyen, and Yingfu Li, "Aptamer-Based Biosensors for Environmental Monitoring," *Frontiers in Chemistry* 8, 1–24, 2020.

236 **aptamers could be useful as therapeutics:** Harleen Kaur, John G. Bruno, Amit Kumar, and Tarun Kumar Sharma, "Aptamers in the Therapeutics and Diagnostics Pipelines," *Theranostics* 8, 4016–32, 2018.

236 **collected and amplified:** Debra L. Robertson and Gerald F. Joyce, "Selection

In Vitro of an RNA Enzyme That Specifically Cleaves Single-Stranded DNA," *Nature* 344, 467–68, 1990.

236 **construct their own nucleotide building blocks:** P. J. Unrau and D. P. Bartel, "RNA-Catalysed Nucleotide Synthesis," *Nature* 395, 260–63, 1998.

236 **act as RNA polymerases:** Wendy K. Johnston, Peter J. Unrau, Michael S. Lawrence, Margaret E. Glasner, and David P. Bartel, "RNA-Catalyzed RNA Polymerization: Accurate and General RNA-Templated Primer Extension," *Science* 292, 1319–25, 2001; David P. Horning and Gerald F. Joyce, "Amplification of RNA by an RNA Polymerase Ribozyme," *Proceedings of the National Academy of Sciences USA* 113, 9786–91, 2016.

236 **attach amino acids to RNA:** Rebecca M. Turk, Nataliya V. Chumachenko, and Michael Yarus, "Multiple Translational Products from a Five-Nucleotide Ribozyme," *Proceedings of the National Academy of Sciences USA* 107, 4585–89, 2010.

INDEX

Abelson, John, 53, 66
aberrant mRNA splicing, 40–43
acridine dyes, 21–23
adaptive immunity, 193, 241, 244
"adaptor molecules." *See* transfer RNA
adenine (A), 10
adenovirus, 31–32, 33
Alberts, Bruce, 119
algae, 228
Alnylam Pharmaceuticals, 162–64, 167, 193
alpha-globin, 41, 241
Alpher, Ralph, 9
ALS (Lou Gehrig's disease), 165, 204
alternative mRNA splicing, 35, 36–37, 42, 241
Altman, Sidney, 62–63, 64, 67
Alzheimer's disease, 166–67, 204, 206
Ambros, Victor, 159
amino acids
   anticodons and, 26
   artificial ribozymes and, 236
   base-substitution errors and, 175

   charges of, 80n
   Covid-19 mRNA vaccines and, 86, 177
   defined, 2, 20, 98, 241
   enzymes and, 98
   formation of, 117
   gene editing and, 218
   genetic code decryption and, 12–13
   hemoglobin and, 41, 219
   mutation and, 145, 177
   peptidyl transfer and, 92
   prolines, 198
   protein chains and, 88, 91, *96*, *105*
   ribosomes and, 102, 104
   sequences of, 14, 25, 34n, 241
   transfer RNA and, 106, 250
   types of, 12
amyloid precursor protein, 167
amyotrophic lateral sclerosis, 206
Anopheles mosquito, 224–25
antibiotics, 102–6
antibodies, 36–37
anticodons, 26, 40, 73–74, *74*, 242
antiphage protection systems, 205–6

INDEX

antisense RNAs
  defined, 42–43, 242
  RNA interference and, 153–55, 168
  therapeutic potential of, 43–44, 203, 204
aplastic anemia, 148
aptamers, 234–36, 242
Archaea, 94
Argonaute, 157–58, *158*, 159, 160, 242
artificial intelligence (AI), 87
Asian mongoose, 226
asthma, 203
Astrachan, Larry, 16
automated sequencers, 95n
Avery, Oswald, 13, 171

bacteriophages
  antiphage protection systems and, 205–6, 206n
  CRISPR DNA and, 207–8
  defined, 15, 242, 246
  intron discovery and, 31
  mRNA discovery and, 16
  number of, 171
  phage T7, 186–88, *187*, 188n, 233
  SunY ribozyme and, 121
  triplet bases and, 21–23
  viral RNA replication and, 176–77
  viral ubiquity and, 170–71
Baker, Chet, 91
Baker, David, 261
Banfield, Jill, 208
Bartel, Dave, 162
base pairs
  complementary, 39, 120, 136, 243
  CRISPR and, 206–7
  defined, 26, 242
  double helix and, 10–11, *10*
  mRNA splicing and, *39*
  ribosomal RNA and, 94
  RNA interference and, 161–62
  RNA structure research and, 72, *73*, 84–86

bases
  decoding and, 19
  defined, 10n, 242
  genetic code decryption and, 12–14
  human genome and, 18
  See also base pairs
base-substitution errors, 175
B cells, 36–37, 203, 242
Berget, Sue, 31
beta-globin, 41, 242
beta-thalassemia, 41, 219
Bethe, Hans, 9
"Big Bang" theory, 1, 9, 231
biochemistry, 23
biofuels, 228
biolistics, 214n
biomedical research, 147, 203–4, 234
BioNTech, 195, 196, 197, 199, 201, 202
Blackburn, Liz, 131–34, 135, 136–37, 144
blood-brain barrier, 166
Bohr, Niels, 9
Boivin, André, 252
Brenner, Sydney, 15, 16, 17, 23, 38, 156–57
Buchner, Eduard, 47–48

cancer
  antisense nucleic acids and, 44
  aptamers and, 235
  death rates, 165
  funding disease-oriented research and, 234
  microRNAs and, 161
  mRNA vaccines for, 5, 185, 195, 197–98, 199, 201–2
  oncogenes and, 206, 246
  telomerase and, 130, 139–40, *140*, 141, 143, 148–51
  TERT promoter mutations and, 149–50, 233
  therapeutic antibodies and, 203
cane toads, 226

INDEX

capsids, 174, 175, 178–80, *179*, 242
Carell, Thomas, 117
Cas9 protein, 206–7, 209–12, *211*, 212, 217
Cas12a protein, 219
Cate, Jamie, 82, 101
C. elegans, 156
cell membranes, 122, 242
Chaires, Jonathan, 92, 93
Chamberlin, Mike, 188
Charpentier, Emmanuelle, 206, 209, 210, 212
chromosomes
  baker's yeast and, 133
  defined, 18n, 242
  enzymes and, 47
  eukaryotic, 51, 52
  human genome and, 18, 34–35, 50, 141, 244
  mRNA and, 191
  RNA and, 2–3, 71
  telomerase discovery and, 134–35, 138
  See also minichromosomes; telomeres
Church, George, 206
Chylinski, Krzysztof, 210
Clemons, Bil, 100
climate change, 5, 227–29
codons, 19, 20–21, 23, 24–25, 26, 34, 40, 243, 250
Cold Spring Harbor Laboratory, 29–30
complementary, 39, 120, 136, 243
conspiracy theories, 185
Counter, Chris, 146
Covid-19 mRNA vaccines
  alternatives to, 197
  apparent speed of development, 4, 185–86
  building blocks for, 196–97
  eteRNA contest and, 86–87
  FDA approval, 189–90, 271
  knowledge derived from, 202
  limitations of, 200

mRNA synthesization and, 188
public suspicion of, 185
Spike protein and, 86, 178, 180, *187*, 188, 196, 198–99
thermal stability and, 86–87
Crick, Francis
  acridine dyes and, 23
  "adaptor molecules" and, 26, *27*
  double-helix discovery, 1, 10–11, 13
  genetic code decryption and, 12
  on mRNA hypothesis, 14–16
  Joan Steitz and, 38
CRISPR (clustered regularly interspaced short palindromic repeats), 205–29, 206n
  antiviral defense origins, 207–9
  Cas9 protein and, 206–7, *211*, 212
  Cas9 system development, 209–12, 217
  climate change mitigation and, 227–29
  Dead Cas9 and, 216–19, 220
  defined, 208, 243
  DNA repair and, 212–16, *215*
  embryonic cells and, 221–22
  enhancements and, 222
  gene drive method, 224–27
  overview of, 205–7
  potential of, 3, 5, 205, 207, 275
  RNA-guided Cas9 protein technology, 209–12
  societal boundaries on, 222–23
  therapeutic applications, 219–21, 277
Crohn's disease, 203
crystallization, 48, 170
Cullis, Pieter, 193
cystic fibrosis, 203, 219
cytoplasm, 13, 14–15, 17, 31–32, 174
cytosine (C), 10

Dahlberg, Jim, 66
dark matter DNA, 231–32
Darwin, Charles, 112
Das, Rhiju, 84, 87, 261
Dead Cas9, 216–19, 220

INDEX

de Lange, Titia, 144
dengue virus, 174
Dicer, 157, 158, *158*, 159, 162, 243
DNA (deoxyribonucleic acid)
  age of, 1–2
  bacterial genome sequences and, 207–8
  base-editing enzymes and, 218, 220
  base sequence of, 25
  CAS9 protein and, 206–7, *211*, 212
  chemical composition of, 13
  dark matter in, 231–32
  deciphering the genetic code and, 12
  double-helical structure of, 1, 3–4, 10–11, *10*, 13, 44, 48, 70
  eukaryotic chromosomes and, 51, 52
  four bases of, 12, 14
  protein synthesis and, 18–19
  repair processes, 212–16, *215*
  replication of, 16, 21–22, 111, 166
  sequences of, 32, 119, 132–34, 243
  telomerase discovery and, 131–34, 135–36, *137*
  tumor sequences and, 149
  See also genome
DNA polymerases, 98, 136
DNA sequences, 32, 119, 132–34, 243
DNA vaccines, 184, 190, 191, 192, 197, 243
DNA viruses, 172, 206–7
donor template, 214, *215*, 216, 217–18, 224, 243
Doudna, Jennifer
  climate crisis research and, 229
  CRISPR and, 206, 208–9, 211–12, 216, 217
  ribozyme structure and, 79, 80–82, 97, 101
  RNA self-replication and, 119–20, 121–22
Dror, Ron, 87
Duchenne muscular dystrophy, 44
Dunn, John, 188
dyskeratosis congenita, 148

Ebola virus, 174
Eckstein, Fritz, 162
E. Coli (*Escherichia coli*)
  aptamers and, 235
  bacteriophage T7 and, 186–88, *187*, 233
  genetic engineering and, 59
  genome of, 18, 175
  intron discovery and, 30
  mRNA discovery and, 17
  phage Q-beta capsid and, 178–79, *179*
  protein synthesis and, 24, 92–93
  ribosomal RNA structure and, 93–94
  RNase P and, 61–62, *62*, 64
  switching phages and, 16
Einstein, Albert, 1
Ellington, Andy, 234–35
Engberg, Jan, 60
English stoats, 226
enzymes
  amino acids and, 98
  Argonaute, 157, 158, *158*, 159, 160, 242
  base-editing, 218, 220
  Cas9 protein, 206–7, 209–12, *211*, 212, 217
  Cas12a protein, 218–19
  chromosomes and, 47
  defined, 47, 243
  Dicer, 157, 158, *158*, 159, 162, 243
  functions of, 47–48
  genetic code decryption, 23–25
  as proteins, 48–49, 55, 61, 62–63, 80, 136
  proteins acting as, 12, 13
  reverse transcriptases as, 98, 136, 144–45, 248
  ribozyme discovery and, 57–61, 64
  RNA circularization and, 56–57
  RNA splicing and, 53–54, *54*, 58–59
  transcription and, 49, 52
  See also ribonuclease; ribozymes; RNA polymerases; telomerase

INDEX

eteRNA contest, 84–87, 261
eugenics, 222
eukaryotes, 32, 33, 34, 50, 213, 243
Euplotes, 142–43, 233
exocytosis, 181, 243

familial disease, 165, 243
Felgner, Phil, 188–89, 190, 192
fermentation, 48, 244
Fire, Andy, 153–55, 156, 157, 162, 168
firefly luciferase, 190–91
flu (influenza), 174, 191–92, 200–201
Fox, George, 94, 95, 262
Fraenkel-Conrat, Heinz, 171–72
frameshift mutation, 22–23, 37

Gall, Joe, 131–33
Gamow, George, 9–10, 11, 12–13, 14, 21, 23, 25
Garraway, Levi, 149–50
gel electrophoresis, 52, 54, 55–56, 60, 63, 160, 244
gene-editing technology. *See* CRISPR
genes, 11
gene therapy, 183–84, 223, 244
genetic code, 25, 26, 33, 102
genetic code decryption, 11, 12–14, 23–25
genetic diseases, 217, 219–21
genetic engineering, 59
genome, 18, 34, 244
  See also human genome
germline cells, 222, 223, 244
Gilboa, Eli, 202
Gold, Larry, 234, 235, 236
Golden, Barb, 83
Gooding, Anne, 81
Grabowski, Paula, 55–57
Gray, Victoria, 221
Green, Michael, 188
greenhouse gases, 228–29
Greider, Carol, 135–38, 139

guanine (G), 10
Guerrier-Takada, Cecilia, 63–64
Guthrie, Christine, 65–67, 258

Harley, Cal, 139
Hayflick, Leonard, 138–39
He Jiankui, 222
HeLa cells, 148
hemoglobin, 33, 41, 219
hepatitis A and, C, 174
hereditary ATTR, 163–64, 193
Hirsh, David, 157
HIV, 98, 144–45
Holley, Robert, 71–72, 73, 253
homologous recombination, 214–16, 215, 217–19, 224, 244
Huang, Franklin, 149–50
Hughes, Tim, 145
human genome
  chromosome size and, 50
  CRISPR DNA and, 207
  frameshift mutation, 22
  introns and, 231
  microRNAs and, 161
  size of, 18, 34–35, 141, 244
  variation in, 237
Human Genome Project (1990–2003), 1, 34–35, 145
human immune system, 184, 193–95, 202, 203, 208, 232, 244
human immunodeficiency virus (HIV), 98, 144–45
human microbiome projects, 95n
human telomerase gene, 137–38, 146

immuno-oncology, 197
influenza (flu), 174, 191–92, 200–201
innate immunity, 193–95, 197, 232, 244
introns
  aberrant mRNA splicing and, 41, 42
  alternative mRNA splicing and, 36, 37
  defined, 244

introns (*continued*)
  discovery of, 30, 31–33, *32*, 39, 44, 51, 76
  human genome and, 231
  snRNAs and, 39–40, 66
  in yeast mitochondria, 76–77
  See also mRNA splicing

Jacob, François, 15, 16, 17, 23
Jansen, Kathrin, 199
Jinek, Martin, 209–12
Joyce, Jerry, 236
"jumping genes," 29

Kahn, Louis, 183
Karikó, Katalin "Kati," 194, 195, 200
Keller, Elizabeth, 24, 72, 73, 75
Khorana, Gobind, 25, 25n
Kim, Sung-Hou, 74
Klug, Aaron, 74
Krainer, Adrian, 42, 43, 155n, 203
kudzu, 226

Lacks, Henrietta, 148
large subunit of the ribosome, *96*, 98–100, 101, 102, 104–5, 244
Larson, Dianne, 43
Larson, Emma, 43
Leder, Phil, 33
Lee, Jeehyung, 261
Lerner, Michael, 38, 40
Lingner, Joachim, 140, 141–42
lipid envelope, 173, 179–80, *180*, 188, 237, 245
lipid nanoparticles (LNP), 189–90, 192–93, *192*, 193, 197, 245
lipids, 122, 189, 245
liposomes, 189, 190, 191, 202, 245
Liu, David, 217–18, 220
long noncoding RNAs (lncRNAs), 232–33, 245
Lou Gehrig's disease (ALS), 165, 204
Lund, Elsebet, 66
Lundblad, Vicki, 144
lymphocytes, 36, 245

machine learning, 87
Madhani, Hiten, 67
malaria, 224–27
Malone, Bob, 183, 184, 185, 189, 190
Maniatis, Tom, 188
Mann, Matthias, 142–43
Martinon, Frédéric, 191–92
Matthaei, Heinrich, 24
Maxwell's equations, 20
McClintock, Barbara, 29, 132
measles, mumps, and rubella (MMR) vaccination, 174
melanoma genome sequences, 149–50
Mello, Craig, 155, 156, 157, 162, 168
Melton, Doug, 188
membrane envelopes, 122–23, 173
Mendel, Gregor, 11, 48
Meselson, Matt, 16, 17
messenger RNA (mRNA)
  Argonaute protein and, 157–58, *158*
  chromosomes and, 191
  defined, 245
  discovery of, 15–17, 23, 253
  human mismatched size, 30–31
  identification of, 23
  negative (-) strand viruses and, 174
  non-vaccine therapies and, 202–4
  physical structure of, 67–68
  positive (+) strand RNA viruses and, 173–74
  ribosome structure and, 101
  ribosome versatility and, 91
  size of, 2
  translocation and, 105, *105*
  See also mRNA splicing; mRNA vaccines; protein synthesis
Meulien, Pierre, 191–92
Meyerson, Matt, 146
Michel, François, 75–79, 83, 84, 95

INDEX

microRNAs, 159–61, 168, 245
military cryptographers, 11
Miller, Stanley, 117
minichromosomes, 131, 133–34, 245
mitochondrial genes, 76–77
Moderna, 195, 196, 199, 201, 202
molecular biology, 33, 155
molecules, 69, 246
Monod, Jacques, 15, 30
Moore, Claire, 31
Moore, Peter, 98
mRNA hypothesis of protein synthesis, 2, 14–17, 101–2
mRNA splicing
   aberrant, 40–43
   alternative, *35*, 36–37, 42, 241
   antisense RNAs and, 43–44
   base pairs and, *39*
   snRNAs and, 39–40, *39*, 44–45, 65–67, 88, 106
mRNA vaccines, 185–202
   alternatives to, 197
   cancer and, 5, 185, 195, 197–98, 199, 201–2
   defined, 246
   for influenza, 200–201
   innate immunity and, 193–95
   lipid delivery vehicles for, 188–93
   mRNA synthesization and, 186–88, *187*
   origins of, 183–85
   potential of, 200, 204
   See also Covid-19 mRNA vaccines
Muller, Hermann, 132
mumps and measles viruses, 174
Murphy, Felicia, 80–81
muscular dystrophy, 203, 219

negative (-) strand RNA viruses, 174, 246
nervous system, 156
NHEJ (non-homologous end joining), 213–14, *215*, 216, 217, 246

Nirenberg, Marshall, 23–24, 25, 91–92
Noller, Harry, 91–96, 100, 104
noncoding RNAs, 45, 232, 246
   See also ribosomal RNA; small nuclear RNAs; transfer RNA
non-homologous end joining (NHEJ), 213–14, *215*, 216, 217, 246
nonnative species introduction, 226
Northrop, John, 170n
nuclear physics, 10
nucleotides
   abbreviations for, 10n
   artificial ribozymes and, 236
   defined, 25, 246
   DNA and RNA polymerases and, 136
   prebiotic formation of, 117–18
   RNA replication and, 176–77
   RNA structure and, 83, 88
   rRNA sequence and, 94–95
   sequences of, 71–72, 73–74, *74*, 76
   start and stop codons and, 21n
   telomerase discovery and, *137*
   transcription and, 52

Omicron variant, 177
oncogenes, 206, 246
Oost, John van der, 208
Orgel, Leslie, 114, 118, 119, 123, 125
origins of life, 109–25
   replication and, 110–11
   ribozymes and, 112–14
   RNA self-replication and, 114–15, *115*, 116–24, 173
   RNA viruses and, 173
orphan diseases, 163
Oxford-AstraZeneca DNA vaccine, 197

Pace, Norm, 67
parasites, 170, 175
Parker, Roy, 66
peptidyl transfer, 92, *96*, *105*, 246
Pfizer Inc, 199

phages. *See* bacteriophages
photosynthesis, 229
physics, 1, 9, 20
plasmid, 59, 247
Platonic solids, 178
poliovirus, 174, 183
polymerase chain reaction (PCR), 235
positive (+) strand RNA virus, 173–74, 247
precision-guided delivery system, 217
precursor mRNA, 33, 37, 42
   See also RNA splicing
precursors, 63–64, 247
   See also precursor mRNA
Prescott, David, 141
prolines, 198
protein-coding genes, 34, 35
proteins
   acting as enzymes, 12, 13
   antibodies and, 36–37
   defined, 247
   diversity of, 12
   enzymes as, 48–49, 55, 61, 62–63, 80, 136
   genetic code decryption and, 12
   human essential, 34
   mutated, 201–3, 204
   origins of life theory and, 124n
   physical structure of, 69–70, 80, 250
   polysaccharide, 254
   RNA interference and, 157–58, 160
   viruses and, 170
   See also amino acids
protein-sequencing methods, 143
protein synthesis
   extracellular, 24
   linguistic analogy for, 17–23
   mRNA hypothesis of, 2, 14–17, 101–2
   ribosomal RNA and, 28, 73, 91–92, 106–7
   ribosome role in, 16–17, 23, 24, 33, 73, 91–92, 98, 101

transfer RNA and, 28, 91, 253
   See also ribozymes
pseudoU, 195, 247, 273
pulmonary fibrosis, 148

rabies virus, 174
Ramakrishnan, Venki, 97, 98, 100, 104
RCSB Protein Data Bank (PDB), 261
receptors, 37, 173, 177, *180*, 247
replicases, 111, 114, 120, 178, 247
replication, 16, 21–22, 110–11, 166, 247
   See also RNA self-replication; viral RNA replication
respiratory syncytial virus (RSV), 174
reverse transcriptases, 98, 136, 144–45, 248
   See also TERT
rheumatoid arthritis, 203
rhinovirus, 174
ribonuclease (RNase)
   capsids and, 178
   defined, 61, 248
   liposomes and, 190
   phage RNA evolution and, 177
   as RNA-degrading enzyme, 136, 163, 195
ribonuclease P (RNase P), 61–64, 137, 248
ribosomal RNA (rRNA)
   antibiotic resistance and, 104
   automated sequencers and, 95n
   base pairs and, 94
   defined, 248
   diagram of, *96*
   introns and, 51, 114n
   mRNA and, 17
   as noncoding RNAs, 45
   protein synthesis and, 28, 73, 91–92, 106–7
   RNA splicing and, *54*, *58*, *78*
   structure research, 92–96, 262
   therapeutic applications, 102–6
   transcription and, 50–52
ribosomes
   amino acids and, 102, 104, *105*

INDEX

bacteria and, 102–3
defined, 5, 248
enzymatic function of, 101–2, 137
protein synthesis and, 16–17, 23, 24, 33, 73, 91–92, 98, 101
pseudoU and, 195
RNA viruses and, 174
structure of, 71, 89, 90–91
structure research, 96–101
subunits of, 96, 98–101, 102, 104–6, 244, 249
transfer RNA and, 250
ribozymes
artificial, 236
defined, 248
discovery of, 4–5, 53–61, 64, 112, 136
"hammerhead," 65
origins of life and, 112–14
ribosomes as, 100
RNase P as, 61–64
RNA structure and, 67–68
snRNAs as, 65–67
structure of, 79–83, 97, 101, 260
synthetic biology and, 89
telomerase as, 136–37, *137*
Rich, Alex, 74
Richardson, Charles, 188
RNA (ribonucleic acid)
about, 2–6, 252
alphabet of, 13–14
circularization, 55–57, 58, 60
future of, 231–37
as life's great catalyst, 71, 236–37
nucleotides of, 116–18
physical structure of, 70–89, *73*
potential of, 3
versatility of, 2–3, *8*
viruses and, 172–73, 177, 178
See also antisense RNAs; messenger RNA; ribosomal RNA; small interfering RNA; small nuclear RNAs; transfer RNA

RNA circularization, 55–57, 58, 60
RNA interference (RNAi), 233, 237
antisense RNA and, 153–55, 168
defined, 248
microRNAs and, 159–61
small interfering RNA discovery, 157–58
therapeutic applications, 156, 161–68
RNA polymerases
defined, 248
long noncoding RNAs and, 232
mRNA synthesization and, 186–88, *187*
mRNA vaccines and, 233
nucleotide synthesizing and, 136
structure of, 98
transcription and, 52, 59
viral replication and, 174, 176
RNA processing, 61, 248
See also RNA splicing
RNA-Puzzles, 88
RNase. *See* ribonuclease
RNA self-replication, 114–15, *115*, 116–24, 173, 193
RNase P. *See* ribonuclease P
RNA sequences
antisense RNA and, 42–43
Covid-19 vaccines and, 86–87
defined, 248
RCSB Protein Data Bank as repository for, 261
ribosomal RNA research and, 94–95
RNA structure research and, 76, 77–78, 84–85, *85*, 261
SARS-CoV-2 virus published, 196, 198
self-replication and, 114n
See also RNA splicing
RNA splicing
ALS genetic cause and, 165–66
defined, 248
enzymes and, 53–54, *54*, 58–59

RNA splicing (*continued*)
  human genome variation and, 237
  inefficiency of, 44
  intron discovery and, 32–33, *32*, 76
  ribosomal RNA and, *54*, *58*, *78*
  ribozyme discovery, 53–55, 57
  self-powered, 60–61, 136
  See also mRNA splicing
RNA structure research, 70–89
  artificial intelligence and, 87–88
  bioengineering applications and, 88–89
  Covid-19 mRNA vaccines and, 86–87
  crowdsourcing and, 84–87, *85*, 261
  drug development and, 83–84
  ribozyme structure and, 79–83, 260
  self-splicing intron model, 75–79, *78*
  tRNA and, 71–75, *73*, *74*
  Watson and, 5, 70–71
"RNA Tie Club," 13, 14, 23
Roberts, Rich, 30, 31, 51, 76
Robertus, J. D., 74
Rosbash, Mike, 258
rRNA. *See* ribosomal RNA
Rubella (German measles), 174
Ruvkun, Gary, 159

Sahin, Ugur, 197–98
Salk, Jonas, 183
Sanger, Fred, 131
SARS-CoV-2 virus
  exocytosis and, 181
  genome proteins and, 175
  inactivated viruses as therapeutic, 197
  lipid envelope of, 173, *180*
  mutations and, 177–78
  origin of life hypothesis and, 173
  RNA as virus genetic material and, 5
  RNA sequence published, 196, 198
  See also Covid-19 mRNA vaccines; Spike protein

Schopf, Bill, 113
senescence, 130, 138, 139, 144, 147, 248
sequencing of nucleic acids
  automated, 95n
  defined, 34n
  See also DNA sequences; RNA sequences
serial transfer experiment, 176–77
Shampay, Janis, 134
SHAPE method, 85, *85*
Sharp, Phil
  Alnylam Pharmaceuticals and, 162, 193
  Andy Fire and, 154
  intron discovery and, 30, 31–33, 51, 76
  mRNA splicing and, 40
  Tom Tuschl and, 160, 162
sickle cell disease, 203, 217, 219, 220–21, 277
single-guide RNA, 212, 248
siRNA. *See* small interfering RNA
Slotta, Karl, 171–72
small interfering RNA (siRNA)
  defined, 249
  Dicer and, 157
  as mRNA-cleaving therapeutics, 193
  neurodegenerative diseases and, 204
  RNA interference and, *158*
  therapeutic applications, 162–68, 204
small nuclear RNAs (snRNAs)
  aberrant mRNA splicing and, 42
  defined, 249
  discovery of, 37–40, 258
  mRNA splicing and, 39–40, *39*, 44–45, 65–67, 88, 106
  as ribozymes, 65–67
small subunit of the ribosome, *96*, 98, 100–101, 102, 105–6, 249
snRNAs. *See* small nuclear RNAs
somatic cells, 221–22, 249
Spiegelman, Sol, 176–77, 178, 234

Spike protein (SARS-CoV-2 virus)
  alternatives to mRNA vaccines and, 197
  Covid-19 mRNA vaccines and, 86, 178, 180, *187*, 188, 196, 198–99
  crowdsourcing the sequencing, 86
  lipid envelope and, 180, *180*
  mutations and, 177–78
  positive (+) strand viruses and, 174
  receptor for, 173
  as red flag, 201
  sequence of, 86, 196
spinal muscular atrophy (SMA), 41–44, 153–54, 155n, 203, 204, 249
splicing. *See* mRNA splicing; RNA splicing
splicing factors, 42, 249
sporadic disease, 165, 249
Stahl, Frank, 16
Stanley, Wendell, 169–70, 171
Stark, Ben, 62
Steitz, Joan Argetsinger, 37–40, 66, 81
Steitz, Tom, 38, 66, 81, 97, 98, 104
stem cells, 130, 147–48, 150, 221, 237
Studier, Bill, 186–88, 195
Sumner, James, 48, 57, 170, 170n
survival of motor neurons (SMN1 and SMN2), 41–42, 249
Sutherland, John, 117
synthetic biology, 89
Szostak, Jack, 119, 121, 123, 133–34, 144, 234–35, 236

Tabor, Stan, 188
tau, 167
Tay-Sachs disease, 219
T cells, 191–92, 191n, 202, 249
telomerase
  cancer and, 130, 139–40, *140*, 141, 143, 148–51
  defined, 250
  discovery of, 130–36, 233
  Hayflick limit and, 139, 146–47
  human telomerase gene, 137–38, 146
  immortality and, 129–30, 139, 147
  purification of, 141–42, *143*
  as RNA-powered enzyme, 5
  senescence prevention and, 138–39, *140*, 146–47
  TERT discovery and, 139–46, *143*
telomeres
  defined, 250
  DNA sequence, 132–34
  Hayflick limit and, 139
  human, 137–38
  hypothesis for aging and, 139, *140*, 143, 147, 148
  senescence and, 148
  telomerase function and, 130
TERT (telomerase reverse transcriptase), 139–46, *143*, 146–47, 149–50, 233, 250
test-tube evolution, 234–36
Tetrahymena thermophila
  ribozyme discovery and, 55–57, 58, *58*, 59–60, 136
  RNA splicing and, 52–53, *54*, 55, *58*, 59
  RNA structure and, 75, 76–79, *78*
  telomerase discovery and, 131–34, 135–36, *137*
  transcription and, 49–52
therapeutic antibodies, 203–4
thymine (T), 10
Tilghman, Shirley, 33, 33n
tobacco mosaic virus (TMV), 169, 170, 172, 174
tracrRNA (trans-activating CRISPR RNA), 210, 211, *211*, 212, 250
transcription, 49–52, 59, 149–50, 195, 250
transfer RNA (tRNA), *27*
  amino acids and, 106, 250
  defined, 250

transfer RNA (tRNA)(*continued*)
  discovery of, 26–27
  as noncoding RNA, 45
  protein synthesis and, 28, 91, 253
  ribonuclease P and, 61, *62*
  ribosomal RNA structure and, 93, *96*
  ribosome structure and, 101
  RNaseP and, 64
  RNA structure research and, 71–75, *73*, *74*, 88
transfer RNA (tRNA) precursors, 63–64
translation, 86, 159–60, 195, 250
translocation, 105, *105*, 106, 250
transthyretin (TTR), 163–64
Treuille, Adrien, 261
triplet codons, 14, 19, 20–23, 24–25
tRNA. *See* transfer RNA
Tuerk, Craig, 234
Türeci, Özlem, 197–98
Tuschl, Tom, 160, 161, 162, 193

Uhlenbeck, Olke, 66, 81, 188
Urey, Harold, 117

vaccines, 184, 191, 192, 197, 243
  See also mRNA vaccines
Verma, Inder, 183, 185, 189
viral RNA replication, 171, 172–78
viruses, 169–81
  capsids, 174, 175, 178–80, *179*, 242
  DNA-based, 172, 206–7
  efficiency of, 181
  intron discovery and, 31
  lipid envelopes, 179–80, *180*
  number of, 170–71
  proteins and, 170
  See also influenza; SARS-CoV-2 virus; viral RNA replication
Volkin, Ken, 16

Watson, James
  Cold Spring Harbor Laboratory and, 29–30
  double-helix discovery, 1, 10–11
  on Gamow, 9
  genetic code decryption and, 12, 13
  intron discovery and, 32
  mRNA discovery and, 253
  mRNA hypothesis and, 14–16
  RNA structure and, 5, 70–71
Weeks, Kevin, 85
Weinberg, Bob, 146
Weissman, Drew, 195, 200
Weissman, Sherman, 41
Westhof, Eric, 77–79, 83, 84, 88
Woese, Carl, 94, 95, 100, 104, 262
Wolff, Jon, 190
World Health Organization, 177, 200
Wright, Woody, 146

X-ray crystallography
  defined, 70, 250
  ribosomal RNA and, 104
  ribosome structure and, 97, 99
  ribozyme structure and, 80
  RNA-Puzzles game and, 88
  RNA structure research and, 70, 74–75, 77, 79, 81–82, 83, 84
Xu, SiQun, 153–55

yeast genome, 34
Yonath, Ada, 97, 98, 104
Yong-Zhen Zhang, 196

Zamecnik, Paul, 24, 26–27
Zamore, Phil, 162
Zaug, Art, 49, 50–54, 59, 160
zebra mussels, 226
Zhang, Feng, 206
zoonosis, 173